Light, Magnetism, and Electricity

At the height of the Age of Newton, all questions seemed answered. But during the next century, as scientists pursued investigations into the realms of light, magnetism, and electricity, it became disturbingly clear that the physical universe was far more complex than Newton and his followers had presumed. In this second volume of his celebrated UNDERSTANDING PHYSICS, Isaac Asimov deals with the collapse of old certainties and the development of bold new theories that have radically altered man's view of the world. Written with marvelous clarity, filled with information, this book represents the perfect supplement to the student's formal textbook, as well as offering invaluable illumination to the general reader.

ABOUT THE AUTHOR: ISAAC ASIMOV is generally regarded as one of this country's leading writers of science and science fiction. He obtained his Ph.D. in chemistry from Columbia University and was Associate Professor of Biochemistry at Boston University School of Medicine. He is the author of over 200 books, including *The Chemicals of Life, The Genetic Code, The Human Body, The Human Brain,* and *The Wellsprings of Life,* all available in Mentor editions.

UNDERSTANDING PHYSICS

Volume II

Light, Magnetism, and Electricity

ISAAC ASIMOV

A MENTOR BOOK

NEW AMERICAN LIBRARY

MENTOR TRADEMARK REG. U.S. PAT. OFF. AND FOREIGN COUNTRIES
REGISTERED TRADEMARK—MARCA REGISTRADA
HECHO EN CHICAGO, U.S.A.

SIGNET, SIGNET CLASSIC, MENTOR, ONYX, PLUME, MERIDIAN AND
NAL BOOKS *are published by NAL PENGUIN INC.,*
1633 Broadway, New York, New York 10019

FIRST MENTOR PRINTING, APRIL, 1969

11   12   13   14   15   16   17   18   19

PRINTED IN THE UNITED STATES OF AMERICA

TABLE OF CONTENTS

CHAPTER 1

Mechanism

The Newtonian View

In the first volume of this book, I dealt with energy in three forms: motion (kinetic energy), sound, and heat. As it turned out, sound and heat are forms of kinetic energy after all. In the case of sound, the atoms and molecules making up the air, or any other medium through which sound travels, move back and forth in an orderly manner. In this way, waves of compression and rarefaction spread out at a fixed velocity (see page I–156).* Heat, on the other hand, is associated with the random movement of the atoms and molecules making up any substance. The greater the average velocity of such movement, the greater the intensity of heat (see page I–234).

By the mid-nineteenth century the Scottish physicist James Clerk Maxwell (1831–1879) and the Austrian physicist Ludwig Boltzmann (1844–1906) had worked out, in strict detail, the interpretation of heat as random molecular movement (the "kinetic theory of heat"). It then became more tempting than ever to suspect that all phenomena in the universe could be analyzed as being based on matter in motion.

* When it is necessary to refer to a passage in the first volume, I will precede the page reference by "I." When the reference is to a page in this volume, it will be given without qualification. In other words, I will say "see page I–123" for a reference to the first volume but "see page 123" for a reference to this one.

According to this view, one might picture the universe as consisting of a vast number of parts; each part, if moving, affecting those neighboring parts with which it makes contact. This is exactly what we see, for instance, in a machine like an ordinary clock. One part of the clock affects another by the force of an expanding spring; by moving, interlocking gears; by levers; in short, by physical interconnections of all kinds. In other machines, such interconnections might consist of endless belts, pulleys, jets of water, and so on. On the submicroscopic scale it is atoms and molecules that are in motion, and these interact by pushing each other when they collide. On the cosmic scale, it is the planets and stars that are in motion, and these interact with each other through gravitational influence.

From the vast universe down to the tiniest components thereof, all might be looked on as obeying the same laws of mechanics by physical interaction as do the familiar machines of everyday life. This is the philosophy of mechanism, or the mechanistic interpretation of the universe. (Gravitational influence does not quite fit this view, as I shall point out shortly.)

The interactions of matter in motion obey, first of all, the three laws of motion (see page I–23ff.) propounded by Isaac Newton (1642–1727) in 1687, and the law of universal gravitation that he also propounded. The mechanistic view of the universe may therefore be spoken of, fairly enough, as the "Newtonian view of the universe."

The entire first volume of this book is devoted to the Newtonian view. It carries matters to the mid-nineteenth century, when this view had overcome all obstacles and had gained strength until it seemed, indeed, triumphant and unshakable.

In the first half of the nineteenth century, for instance, it had been found that Uranus traveled in its orbit in a way that could not be quite accounted for by Newton's law of universal gravitation. The discrepancy between Uranus's actual position in the 1840's and the one it was expected to have was tiny; nevertheless the mere existence of that discrepancy threatened to destroy the Newtonian fabric.

Two young astronomers, the Englishman John Couch Adams (1819–1892) and the Frenchman Urbain Jean Joseph Leverrier (1811–1877), felt that the Newtonian view could not be wrong. The discrepancy had to be due to the existence of an unknown planet whose gravitational influence on Uranus was not being allowed for. Independently they calculated where such a planet had to be located to account for the observed discrepancy in

Uranus's motions, and reached about the same conclusion. In 1846 the postulated planet was searched for and found.

After such a victory, who could doubt the usefulness of the Newtonian view of the universe?

And yet, by the end of the century, the Newtonian view had been found to be merely an approximation. The universe was more complicated than it seemed. Broader and subtler explanations for its workings had to be found.

Action at a Distance

The beginnings of the collapse were already clearly in view during the very mid-nineteenth-century peak of Newtonianism. At least, the beginnings are clearly to be seen by us, a century later, with the advantage of hindsight. The serpent in the Newtonian Eden was something called "action at a distance."

If we consider matter in motion in the ordinary world about us, trying to penetrate neither up into the cosmically vast nor down into the submicroscopically small, it would seem that bodies interact by making contact. If you want to lift a boulder you must touch it with your arms or use a lever, one end of which touches the boulder while the other end touches your arms.

To be sure, if you set a ball to rolling along the ground, it continues moving even after your arm no longer touches it; and this presented difficulties to the philosophers of ancient and medieval times. The Newtonian first law of motion removed the difficulty by assuming that only *changes* in velocity required the presence of a force (see page I-24). If the rolling ball is to increase its velocity, it must be struck by a mallet, a foot, some object; it must make contact with something material. (Even rocket exhaust, driving backward and pushing the ball forward by Newton's third law of motion, makes material contact with the ball.) Again, the rolling ball can be slowed by the friction of the ground it rolls on and touches, by the resistance of the air it rolls through and touches, or by the interposition of a blocking piece of matter that it must touch.

Material contact can be carried from one place to another by matter in motion. I can stand at one end of the room and knock over a milk bottle at the other end by throwing a ball at it. I exert a force on the ball while making contact with it; then the ball exerts a force on the bottle while making contact with it. We have two contacts connected by motion. If the milk bottle is balanced precariously enough, I can knock it over by blowing at it. In that case, I throw air molecules at it, rather than a ball, but the principle is the same.

Is it possible, then, for two bodies to interact without physical contact at all? In other words, can two bodies interact across a vacuum without any material bodies crossing that vacuum? Such action at a distance is very difficult to imagine; it is easy to feel it to be a manifest impossibility.

The ancient Greek philosopher Aristotle (384–322 B.C.), for instance, divined the nature of sound partly through a refusal to accept the possibility of action at a distance. Aristotle felt that one heard sounds across a gap of air because the vibrating object struck the neighboring portion of air, and that this portion of the air passed on the strike to the next portion, the process continuing until finally the ear was struck by the portion of the air next to itself. This is, roughly speaking, what does happen when sound waves travel through air or any other conducting medium. On the basis of such an interpretation, Aristotle maintained that sound could not travel through a vacuum. In his day mankind had no means of forming a vacuum, but two thousand years later, when it became possible to produce fairly good vacuums, Aristotle was found to be correct.

It might follow, by similar arguments, that all interactions that seem to be at a distance really consist of a series of subtle contacts, and that no interaction of any kind can take place across a vacuum. Until the seventeenth century it was strongly believed that a vacuum did not exist in nature but was merely a philosophical abstraction, so there seemed no way of testing this argument.

In the 1640's, however, it became clear that the atmosphere could not be infinitely high (see page I–146). Indeed, it was possibly no more than a few dozen miles high, whereas the moon was a quarter of a million miles away, and other astronomical bodies were much farther still. Any interactions between the various astronomical bodies must therefore take place across vast stretches of vacuum.

One such interaction was at once obvious, for light reaches us from the sun, which we now know is 93,000,000 miles away.* This light can affect the retina of the eye. It can affect the chemical reactions proceeding in plant tissue; converted to heat, it can evaporate water and produce rain, warm air, and winds. Indeed, sunlight is the source of virtually all energy used by man. There is

* Our best telescopes can detect light that has traversed some 35,000,000,-000,000,000,000 miles of vacuum.

thus a great deal of interaction, by light, between the sun and the earth across the vast vacuum.

Then, once Newton announced the law of universal gravitation in 1687, a second type of interaction was added, for each heavenly body was now understood to exert a gravitational force on all other bodies in the universe across endless stretches of vacuum. Where two bodies are relatively close, as are the earth and the moon or the earth and the sun, the gravitational force is large indeed, and the two bodies are forced into a curved path about their common center of gravity. If one body is much larger than the other, this common center of gravity is virtually at the center of the larger body, which the smaller then circles.

On the earth itself, two additional ways of transmitting force across a vacuum were known. A magnet could draw iron to itself, and an electrically charged body could draw almost any light material to itself. One magnet could either attract or repel another; one electric charge could either attract or repel another. These attractions and repulsions could all be exerted freely across the best vacuum that could be produced.

In the mid-nineteenth century, then, four ways of transmitting force across a vacuum, and hence four possible varieties of action at a distance, were known: light, gravity, electricity, and magnetism. And yet the notion of action at a distance was as unacceptable to nineteenth-century physicists as it had been to the philosophers of ancient Greece.

There were two possible ways out of the dilemma; two ways of avoiding action at a distance.

First, perhaps a vacuum was not really a vacuum. Quite clearly a good vacuum contained so little ordinary matter that this matter could be ignored. But perhaps ordinary matter was not the only form of substance that could exist.

Aristotle had suggested that the substance of the universe, outside the earth itself, was made up of something he called *ether*. The ether was retained in modern science even when virtually all other portions of Aristotelian physics had been found wanting and had been discarded. It was retained, however, in more sophisticated fashion. It made up the fabric of space, filling all that was considered vacuum and, moreover, permeating into the innermost recesses of all ordinary matter.

Newton had refused to commit himself as to how gravitation was transmitted from body to body across the void. "I make no hypotheses," he had said austerely. His followers, however, pictured gravitation as making its way through the ether much as sound makes its way through air. The gravitational effect of a body would be expressed as a distortion of that part of the ether with which it made contact; this distortion would right itself and; in the process, distort a neighboring portion of the ether. The traveling distortion would eventually reach another body and affect it. We can think of that traveling distortion as an "ether wave."

The second way out of the dilemma of action at a distance was to assume that forces that made themselves felt across a vacuum were actually crossing in the form of tiny projectiles. The projectiles might well be far too small to see, but they were there. Light, for instance, might consist of speeding particles that crossed the vacuum. In passing from the sun to the earth, they would make contact first with the sun and then with the earth, and there would be no true action at a distance at all, any more than in the case of a ball being thrown at a bottle.

For two centuries after Newton, physicists vacillated between these two points of view: waves and particles. The former required an ether, the latter did not. This volume will be devoted, in good part, to the details of this vacillation between the two views. In the eighteenth century, the particle view was dominant; in the nineteenth, the wave view. Then, as the twentieth century opened, a curious thing happened—the two views melted into each other and became one!

To explain how this happened, let's begin with the first entity known to be capable of crossing a vacuum—light.

CHAPTER 2

Light

Transmission

Surely light broke in on man's consciousness as soon as he had any consciousness at all. The origins of the word itself are buried deep in the mists of the beginnings of the Indo-European languages. The importance of light was recognized by the earliest thinkers. In the Bible itself, God's first command in constructing an ordered universe was "Let there be light!"

Light travels in straight lines. This, indeed, is the assumption each of us makes from babyhood. We are serenely sure that if we are looking at an object that object exists in the direction in which we are looking. (This is strictly true only if we are not looking at a mirror or through a glass prism, but it is not difficult to learn to make the necessary exceptions to the general rule.)

This straight-line motion of light, its *rectilinear propagation*, is the basic assumption of *optics* (from a Greek word meaning "sight"), the study of the physics of light. Where the behavior of light is analyzed by allowing straight lines to represent the path of light and where these lines are studied by the methods of geometry, we have *geometric optics*. It is with geometric optics that this chapter and the next are concerned.

Consider a source of light such as a candle flame. Assuming that no material object blocks your vision at any point, the flame can be seen with equal ease from any direction. Light, therefore, can be visualized as streaming out from its source in all directions. The sun, for instance, can be drawn (in two dimensions) as a circle with lines, representing light, extending outward from all parts of the circumference.

Such lines about the drawing of the sun resemble spokes of a wheel emerging from the hub. The Latin word for the spoke of a wheel is *radius* (which gives us the word for the straight line extending from the center of a circle to its circumference). For this reason, the sun (or any light source) is said to *radiate* light, and light is spoken of as a *radiation*. A very thin portion of such a light radiation, one that resembles a line in its straightness and ultimate thinness, is a *light ray*, again from *radius*.

Sunlight shining through a hole in a curtain will form a pillar of light extending from the hole to the opposite wall where the intersection of the pillar with the wall will form a circle of bright illumination. If the air of the room is normally dusty, the pillar of light will be outlined in glittering dust motes. The straight lines bounding the pillar of light will be visible evidence of the rectilinear propagation of light. Such a pillar of light is a *light beam* (from the resemblance of its shape to the trunk of a tree; the German word for tree is "Baum," and a similar word, of course, is found in Anglo-Saxon). A light beam may be viewed as a collection of an infinite number of infinitesimally thin light rays.

Light sources vary in brightness. More light is given off by a hundred-watt light bulb than by a candle, and incomparably more light still is given off by the sun. To measure the quantity of light given off by a light source, physicists can agree to use some particular light source as standard. The obvious early choice for the standard was a candle made of a specified material (sperm wax was best) prepared in a particular way and molded to set specifica-

Variation of light intensity with distance

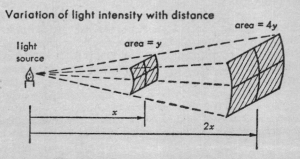

light source

area = y

area = $4y$

x

$2x$

tions. The light emitted by this candle horizontally could then be said to equal 1 *candlepower.* Electric light bulbs of set form have now replaced the candle, especially in the United States, but we still speak of the *international candle,* a measure of light quantity about equal to the older candlepower.

The brightness of a light source varies in some fashion with the distance from which it is viewed: the greater the distance, the dimmer it seems. A book held near a candle may be read easily; held farther away it becomes first difficult and then impossible to read.

This is not surprising. Suppose a fixed amount of light is emerging from the candle flame. As it spreads out in all directions, that fixed amount must be stretched over a larger and larger area. We can imagine the edge of the illumination to be forming a sphere with the light source as center. The sphere's surface grows larger and larger as the light radiates outward.

From plane geometry we know that the surface of a sphere has an area proportional to the square of the length of its radius. If the distance from the light source (the radius of the imaginary sphere we are considering) is doubled, the surface over which the light is spread is increased two times two, or 4 times. If the distance is tripled, the surface is increased 9 times. The total quantity of light over the entire surface may remain the same, but the intensity of light—that is, the amount of light falling on a particular area of surface—must decrease. More, it must decrease as the square of the distance from the light source. Doubling the distance from the light source decreases the light intensity to 1/4 the original; tripling the distance decreases it to 1/9.

Suppose we use the square foot as the unit of surface area and imagine that square foot bent into the shape of a segment of a spherical surface so that all parts of it are equidistant from the centrally located light source. If such a square foot is just one foot distant from a light source delivering 1 candle of light, then the intensity of illumination received by the surface is 1 *foot-candle.* If the surface is removed to a distance of two feet, the intensity of its illumination is 1/4 foot-candle, and so on.

Since light intensity is defined as the quantity of light per unit area, we can also express it as so many candles per square foot. For this purpose, however, a unit of light quantity smaller than the candle is commonly used. This is the *lumen* (from a Latin word for "light"). Thus if one square foot at a certain distance from a light source receives 1 lumen of light, two square feet at that same distance will receive 2 lumens of light, and half a square foot will receive 1 2 lumen. In each case, though, the light intensity will be 1 lumen/foot.^{2} The lumen is so defined that an intensity of 1 lumen/foot^2 equals 1 foot-candle.

Imagine a light source of 1 candle at the center of a hollow sphere with a radius of one foot. The light intensity on each portion of the interior surface of the sphere is 1 foot-candle, or 1 lumen/foot.^{2} Each square foot of the interior surface is therefore receiving 1 lumen of illumination. The area of the surface of the sphere is equal to $4\pi r^2$ square feet. Since the value of r, the radius of the sphere, is set at 1 foot, the number of square feet of surface equals 4π. The quantity π (the Greek letter *pi*) is equal to about 3.14, so we can say that there are about 12.56 square feet on that spherical surface. The light (which we have set at 1 candle) is therefore delivering a total of 12.56 lumens, so we can say that 1 candle equals 12.56 lumens.

Light is transmitted, completely and without impediment, only through a vacuum. All forms of matter will, to some extent at least, absorb light. Most forms do so to such an extent that in ordinary thicknesses they absorb all the light that falls on them and are *opaque* (from a Latin word meaning "dark").

If an opaque object is brought between a light source and an illuminated surface, light will pass by the edges of the object but not through it. On the side of the object opposite the light source there will therefore be a volume of darkness called a *shadow*. Where this volume intersects the illuminated surface there will be a non-illuminated patch; it is this two-dimensional intersection of the shadow that we usually refer to by the word.

The moon casts a shadow. Half its surface is exposed to the direct illumination of the sun; the other half is so situated that the opaque substance of the moon itself blocks the sunlight. We see only the illuminated side of the moon, and because this illuminated side is presented to us at an angle that varies from 0° to 360° during a month, we watch the moon go through a cycle of phases in that month.

Furthermore, the moon's shadow not only affects its own surface, but stretches out into space for over two hundred thousand miles. If the sun were a "point source"—that is, if all the light came from a single glowing point—the shadow would stretch out indefinitely. However, the sun is seen as an area of light, and as one recedes from the moon its apparent size decreases until it can no longer cover all the area of the much larger sun. At that point, it no longer casts a complete shadow, and the complete shadow (or *umbra,* from a Latin word for "shadow") narrows to a point. The umbra is just long enough to reach the earth's surface, however, and on occasion, when the moon interposes itself exactly between earth and sun, a *solar eclipse* takes place over a small area of the earth's surface.

The earth casts a shadow, too, and half its surface is in that shadow. Since the earth rotates in twenty-four hours, each of us experiences this shadow ("night") during each 24-hour passage. (This is not always true for polar areas, for reasons better discussed in a book on astronomy.) The moon can pass through the earth's shadow, which is much longer and wider than that of the moon, and we can then observe a *lunar eclipse.*

Opaque materials are not absolutely opaque. If made thin enough, some light will pass through. Fine gold leaf, for instance, will be traversed by light even though gold itself is certainly opaque.

Some forms of matter absorb so little light (per unit thickness) that the thicknesses we ordinarily encounter do not seriously interfere with the transmission of light. Such forms of matter are *transparent* (from Latin words meaning "to be seen across"). Air itself is the best example of transparent matter. It is so transparent that we are scarcely aware of its existence, since we see objects through it as if there were no obstacle at all. Almost all gases are transparent. Numerous liquids, notably water, are also transparent.

It is among solids that transparency is very much the exception. Quartz is one of the few naturally occurring solids that display the property, and the astonished Greeks considered it a form of warm ice. The word "crystal," first applied to quartz, is from their word for "ice," and the word "crystalline" has as one of its meanings "transparent."

Transparency becomes less pronounced when thicker and thicker sections of ordinarily transparent substances are considered. A small quantity of water is certainly transparent, and the pebbles at the bottom of a clear pool can be seen distinctly. However, as a diver sinks beneath the surface of the sea, the light that can reach him grows feebler and feebler, and below about 450 feet almost no light can penetrate. Thicknesses of water greater than that are as opaque as if they represented the same thickness of rock, and the depths of the sea cannot be seen through the "transparent" water that overlays it.

Air absorbs light to a lesser extent than water does and is therefore more transparent. Even though we are at the bottom of an ocean of air many miles high, sunlight has no trouble penetrating to us, and we in turn have no trouble seeing the much feebler light of the stars.* Nevertheless some absorption exists: it is estimated, for instance, that 30 percent of the light reaching us from space is absorbed by that atmosphere. (Some forms of radiation other than visible light are absorbed with much greater efficiency by the atmosphere, and the thickness of air that blankets us suffices to make the air opaque to these radiations.)

Light is a form of energy, and while it can easily be changed into other forms of energy, it cannot be destroyed. While absorption by an opaque material (or a sufficient thickness of a transparent material) seems to destroy it, actually it is converted into heat.

Reflection

The statement that light always travels in a straight line is completely true only under certain circumstances, as when light travels through a uniform medium—through a vacuum, for instance, or through air that is at equal temperature and density throughout. If the medium changes—as when light traveling through air strikes an opaque body—the straight-line rule no longer holds strictly. Such light as is not absorbed by the body changes direction abruptly, as a billiard ball will when it strikes the edge of a pool table.

This bouncing back of light from an opaque body is called *reflection* (from Latin words meaning "to bend back").

The reflection of light seems to follow closely the rules that govern the bouncing of a billiard ball. Imagine a flat surface capable of reflecting light. A line perpendicular to that surface is called the *normal,* from the Latin name for a carpenter's square used to draw perpendiculars.† A ray of light moving along the normal strikes the reflecting surface head-on and doubles back in its tracks. A speeding billiard ball would do the same.

If the ray of light were traveling obliquely with respect to the reflecting surface, it would strike at an angle to the normal. The

---

* To be sure, if the atmosphere were compressed to the density of water, it would be only some 33 feet thick; and that thickness of water would retain considerable transparency, too.

† Straightforward behavior that is "square" and "on the beam," like a perpendicular, accurately drawn by a carpenter's square, is also "normal." Other types of behavior are "abnormal" or represent "enormities." In fact, the word "normal" has become so familiar in its sense of natural, commonplace, conforming behavior that its original meaning of "a line perpendicular to a plane, or to another line" has almost been forgotten.

light ray moving toward the surface is the *incident ray*, and its angle to the normal is the *angle of incidence*. The *reflected ray* would return on the other side of the normal, making a new angle, the *angle of reflection*. The incident ray, reflected ray, and normal are all in the same plane—that is, a flat sheet could be made to pass through all three simultaneously without its flatness being distorted.

Experiments with rays of light and reflecting surfaces in dusty air, which illuminates the light rays and makes them visible, will show that the angle of incidence (*i*) always equals the angle or reflection (*r*). This can be expressed, simply:

$$i = r$$ (Equation 2–1)

Actually, it is rare to find a truly flat surface. Most surfaces have small unevennesses even when they appear flat. A beam of light, made up of parallel rays, would not display the same angle of incidence throughout. One ray might strike the surface at a spot where the angle of incidence is 0°; another might strike very close by where the surface has nevertheless curved until it is at an angle of 10° to the light; elsewhere it is 10° in the other direction, or 20°, and so on. The result is that an incident beam of light with rays parallel will be broken up on reflection, with the reflected rays

Reflection of light

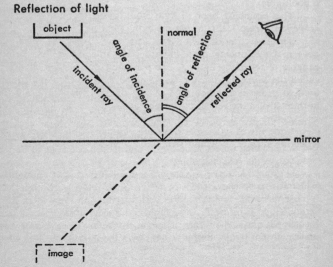

traveling in all directions over a wide arc. This is *diffuse reflection.*

Almost all reflection we come across is of this type. A surface that reflects light diffusely can be seen equally well from different angles, since at each of the various angles numerous rays of light are traveling from the object to the eye.

If a surface is quite flat, a good portion of the parallel rays of incident light will be reflected at the same angle. In such a case, although you can see the reflecting object from various angles, you will see far more light if you orient yourself at the proper angle to receive the main reflection. At that point you will see a "highlight."

If a surface is extremely flat, virtually all the parallel rays of an incident beam of light will be reflected still parallel. As a result, your eyes will interpret the reflected beam as they would the original.

For instance, the rays of light reflected diffusely from a person's face make a pattern that the eyes transmit and the brain interprets as that person's face. If those rays strike an extremely flat surface, are reflected without mutual distortion, and then strike your eyes, you will still interpret the light as representing that person's face.

Your eyes cannot, however, tell the history of the light that reaches them. They cannot, without independent information, tell whether the light has been reflected or not. Since you are used from earliest life to interpreting light as traveling in straight, uninterrupted lines, you do so now, too. The person's face as seen by reflected light is seen as if it were behind the surface of reflection, where it would be if the light had come straight at you without interruption, instead of striking the mirror and being reflected to you.

The face that you see in a mirror is an *image.* Because it does not really exist in the place you seem to see it (look behind the mirror and it is not there) it is a *virtual image.* (It possesses the "virtues" or properties of an object without that object actually being there.) It is, however, at the same distance behind the mirror that the reflected object is before it, and therefore seems to be the same size as the reflected object.

In primitive times virtually the only surface flat enough to reflect an image was a sheet of water. Such images are imperfect because the water is rarely quite undisturbed, and even when it is, so much light is transmitted by the water and so little reflected that the image is dim and obscure. Under such circumstances a primitive man might not realize that it was his own face staring back at him.

(Consider the Greek myth of Narcissus, who fell hopelessly in love with his own reflection in the water and drowned trying to reach it.)

A polished metal surface will reflect much more light, and metal surfaces were used throughout ancient and medieval times as mirrors. Such surfaces, however, are easily scratched and marred. About the seventeenth century the glass-metal combination became common. Here a thin layer of metal is spattered onto a sheet of flat glass. If we look at the glass side, we see a bright reflection from the metal surface covering the other side. The glass serves to protect the metal surface from damage. This is a *mirror* (from a Latin word meaning "to look at with astonishment," which well expresses primitive feelings about images of one's self) or *looking glass*. A Latin word for mirror is *speculum*, and for that reason the phrase for the undisturbed reflection from an extremely flat surface is *specular reflection*.

An image as seen in a mirror is not identical with the object reflected.

Suppose you are facing a friend. His right side is to your left; his left side is to your right. If you want to shake hands, right hand with right hand, your hands make a diagonal line between your bodies. If you both part your hair on the left side, you see his part on the side opposite that of your own.

Now imagine your friend moving behind you but a little to one side so that you can both be seen in the mirror before you. Ignore your own image and consider your friend's only. You are now facing, not your friend, but the image of your friend, and there is a change. His right side is on your right and his left side is on your left. Now the parts in your hair are on the same side, and if you hold out your right hand while your friend holds out his, your outstretched hand and that of the image will be on the same side.

In short, the image reverses right and left; an image with such a reversal is a *mirror image*. A mirror image does not, however, reverse up and down. If your friend is standing upright, his image will be upright, too.

Curved Mirrors

The ordinary mirror with which we are familiar is a *plane mirror*—that is, it is perfectly flat. A reflecting surface, however, need not be flat to exhibit specular reflection. It can be curved, as long as it is smooth. Parallel rays of light reflected from a curved surface are no longer parallel, but neither are they reflected in random directions. The reflection is orderly and the rays of light may *converge* (from Latin words meaning "to lean together") or *diverge* ("to lean apart").

The simplest curvature is that of a section of a sphere. If you are looking at the outside of the section, so that it forms a hill toward you with the center closest to you, it is a *convex surface* (from Latin words meaning "drawn together"). If you are looking at the inside of the spherical section, you are looking into a hollow with the center farthest from you. That is a *concave surface* ("with a hollow").

A spherical segment of glass, properly silvered, is a *spherical mirror*. If it is silvered on its convex surface so that you see it as a mirror if you look into its concave surface, it is, of course, a concave spherical mirror. The center of the sphere of which the curved mirror is part is the *center of curvature*. A line connecting the center of curvature with the midpoint of the mirror is the *principal axis* of the mirror.

Suppose a beam of light, parallel to the principal axis, falls upon the concave reflecting surface. The ray that happens to lie on the principal axis itself strikes perpendicularly and is reflected back upon itself. With a ray of light that strikes near the principal axis but not on it, the mirror has curved in such a way that the ray makes a small angle with the normal. It is reflected on the other side of the normal in a fashion that bends it slightly toward the principal axis. If the ray of light strikes farther from the principal axis, the mirror has bent through a larger angle and reflects the ray more sharply toward the principal axis. Since the mirror is a spherical segment and curves equally in all directions from the principal axis, this is true of rays of light striking either right or left of the principal axis, either above or below it. Reflec-

Concave spherical mirror

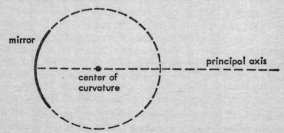

tions from every part of the mirror point toward the principal axis; the reflected rays converge.

If only those rays that strike fairly close to the midpoint of the mirror are considered, it is found that they converge in such a way as to meet in a restricted region—approximately at a point, in fact. This point is called a *focus* (from a Latin word for "hearth," which is where one would expect a concentration of light). The focus falls on the principal axis, halfway between the midpoint of the mirror and the center of curvature.

Actually, the reflected rays do not all meet exactly at the focus. This becomes obvious if we consider rays that fall on the spherical mirror quite a distance from the principal axis. The reflections of these rays miss the focus by a considerable distance. This is called *spherical aberration* (from the Latin, "to wander away"). These distant rays fall between the focus and the mirror itself and are therefore reflected through too great an angle. The mirror, in other words, has curved too sharply to bring all the rays to a focus.

To avoid this, we need a curved mirror that curves somewhat less sharply than a spherical segment does. The necessary curve is that of a *paraboloid of revolution*.

A spherical section, if it is continued, closes in upon itself and finally forms a sphere. A paraboloid of revolution looks like a spherical segment if only a small piece about the midpoint is taken. If it is continued and made larger, it does not close in upon itself. It curves more and more gently till its walls are almost straight, forming a long cylinder that becomes wider only very slowly. A mirror formed of a section (about the midpoint) of such a paraboloid of revolution is called a *parabolic mirror*.

If a beam of light parallel to the principal axis of such a parabolic mirror falls upon its concave surface, the rays do indeed converge upon a focus, and without aberration.

To produce such a beam of light, consisting of parallel rays, we must, strictly speaking, think of a point source of light on the principal axis an infinite distance from the mirror. If the point source is a finite distance away, then the rays striking the mirror from that point source are not truly parallel, but diverge slightly. Each ray strikes the mirror surface at an angle to the normal which is slightly smaller than it would be if the rays were truly parallel, and in consequence is reflected through a smaller angle. The rays therefore converge farther away from the mirror than at the focus. If the distance of the point source is large compared to the distance of the focus (which is only a matter of a few inches for the average parabolic mirror), then the rays converge at a point very near the focus—near enough so that the difference can be ignored.

If the light source is moved closer and closer to the mirror, the reflected rays converge farther and farther from the mirror. When the light source is at twice the distance of the focus from the mirror, eventually the reflected rays converge at the light source itself. If the light source is moved still closer, the reflected rays converge at a point beyond the light source.

Finally, if the light source is located at the focus itself, the reflected rays no longer converge at all, but are parallel. (We might say that the point of convergence has moved an infinite distance away from the mirror.) The automobile headlight works in this fashion. Its inner surface is a parabolic mirror, and the small incandescent bulb is at its focus. Consequently, such a headlight casts a fairly straight beam of light forward.

Let us call the distance of the light source from the mirror D_o, and the distance of the point of convergence of the reflected rays from the mirror, D_1. The distance of the focus from the mirror we can call f. The following relationship then holds true:

$$\frac{1}{D_o} + \frac{1}{D_1} = \frac{1}{f}$$

(Equation 2–2)

We can check this for the cases we have already discussed. Suppose that the light source is at a very great distance (practically infinite). In that case, D_o is extremely large and $1/D_o$ is extremely small. In fact, $1/D_o$ can be considered zero. In that case, Equation

Parabolic mirror

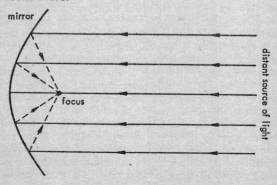

mirror

focus

distant source of light

2–2 becomes $1/D_i = 1/f$, and $D_i = f$, which means that the re-flected rays of light converge at the focus.

If the light source is on the principal axis but at twice the distance from the mirror that the focus is, then $D_o = 2f$, and Equation 2–2 becomes: $1/2f + 1/D_i = 1/f$. If we solve this equation for D_i, we find that $D_i = f$. In other words, the re-flected rays in this case converge upon the location of the light source itself.

And what if the light source is located at the focus? In that case $D_o = f$. Equation 2–2 becomes $1/f + 1/D_i = 1/f$, from which you can see at once that $1/D_i = 0$. But if $1/D_i = 0$, then D_i must be infinitely large. The distance from the mirror at which the reflected rays converge is infinite, and therefore the rays do not converge at all but are parallel.

In the previous section I have been considering the source of light to be a point. Actually, of course, it is not really a point. Suppose the source of light is a candle flame which, naturally, covers an area. Some of the flame is slightly above the principal axis, some slightly below, some to one side, and some to another. The rays of light that originate somewhat above the principal axis are reflected to a point somewhat below the true point of convergence (that is, what would have been the true point if the candle flame had been a point source of light); those that originate below the principal axis are reflected to a point above the point of convergence; those that originate to the right are reflected to the left; those that originate to the left are reflected to the right. If we take any particular ray, the greater the distance from the principal axis it originates, the greater the distance from the point of convergence, but on the opposite side.

The result is that in the area where the reflected rays of light converge, one obtains an image in which not only left and right are interchanged (as in a plane mirror), but also up and down. An upside-down image is formed; indeed, if you look into the shiny bowl of a spoon, you will see your face upside down.

The image produced by such a concave mirror has another important difference from that produced by a plane mirror. The image produced by the plane mirror, as was stated earlier, is not actually behind the mirror where it seems to be, so it is a virtual image. In the case of a concave mirror, the image is formed in front of the mirror by means of converging light rays. The image is really there and can be touched; therefore it is a *real image*.

To be sure, when you actually touch a real image you don't seem to be touching anything, because you are used to considering touch only in connection with matter. A parabolic mirror does not converge matter; it converges light and you cannot touch light in the ordinary sense. However, you can sense light when it is absorbed by the skin and turned to heat; and in that respect, by feeling heat, you are "touching" the image.

A finger held six feet from a candle flame absorbs some heat from the radiation that falls directly upon it. The finger, however, intercepts but a small fraction of the total radiation of the candle, and the heating effect is insignificant. A concave mirror would intercept more of the candle's radiation and converge it to a small volume of space. The finger placed at the area of convergence would feel more heat in that area than elsewhere in the neighborhood. The increase in heat concentration may still be too little to feel, but if the concave mirror is used to concentrate the rays of the sun instead, you will certainly feel it. Large parabolic mirrors have been built that intercept solar radiation over a sizable area and converge it all. Temperatures as high as 7000°C have been reached at the focus of such *solar furnaces*. There is a real image that can be felt with a vengeance.

A mirror of changing curvature can produce odd and humorous distortions in the image, as anyone attending amusement parks knows. However, a proper image from a clean mirror of undistorted shape can seem completely legitimate, particularly if the boundaries of the mirror are masked so that the onlooker has no reason to feel that a mirror is there at all. The casual viewer mistakes image for reality and this is the basis for some of the tricks of the magician. Naturally, a real image is even more tantalizing than a virtual image. At the Boston Museum of Science, a real image is projected in such a way as to make coins seem to tumble about in an upside-down goblet in defiance of gravity. Onlookers (adults as well as children) never tire of placing their hands where the coins seem to be. Not all their insubstantiality can convince the eyes that the coins are not there.

Suppose the light source is moved still closer to the mirror than the distance of the focus. In that case, the reflected rays are neither convergent nor parallel; they actually diverge. Such diverging rays, spraying outward from a surface, may be considered as converging if you follow them backward. Indeed, if you follow them (in imagination) through the mirror's surface and into the space behind, they will converge to a point. At that point you will see an image. Because it appears behind the mirror where the light really doesn't penetrate, it is a virtual image, as in the case of a plane mirror; and, as in the case of a plane mirror, the image is now right-side up.

Equation 2-2 can be made to apply to this situation. If the light source is closer to the mirror than the focus is, then D_o is smaller than f, and $1/D_o$ must therefore be larger than $1/f$. (If this is not at once clear to you, recall that 2 is smaller than 4, and that $1/2$ is therefore larger than $1/4$.)

If we solve Equation 2-2 for $1/D_i$ we find that:

$$\frac{1}{D_i} = \frac{1}{f} - \frac{1}{D_o} \qquad \text{(Equation 2-3)}$$

Since in the case under consideration, $1/D_o$ is larger than $1/f$, $1/D_i$ must be a negative number. From this it follows that D_i itself must be a negative number.

This makes sense. In the previous cases under discussion, distances have all extended forward from the mirror. In the present case the point at which the reflected rays converge, and where the image exists, lies behind the mirror, and its distance should, reasonably, be indicated by a negative value.

Nor need Equation 2-2 be applied only to concave mirrors; it is more general than that.

Consider a plane mirror again. A beam of parallel rays striking it along its principal axis (any line normal to the plane mirror can be considered a principal axis) is reflected back along the principal axis as parallel as ever. The rays do not converge and therefore the distance of the focus from the mirror is infinitely great. But if f is infinitely great, then $1/f$ must equal zero and, for a plane mirror, Equation 2-2 becomes:

$$\frac{1}{D_o} + \frac{1}{D_i} = 0 \qquad \text{(Equation 2-4)}$$

If Equation 2-4 is solved for D_i, it turns out that $D_i = -D_o$. Because D_o (the distance of the object being reflected) must always be positive since it must always lie before the mirror in order to be reflected at all, D_i must always be negative. In a plane mirror, therefore, the image must always lie behind the mirror and be a virtual one. Since, except for sign, D_i and $-D_o$ are equal, the image is as far behind the mirror as the object being reflected is in front of the mirror.

What, now, if we have a *convex mirror*—that is, a curving mirror which is silvered on the concave side so that we look into and see a reflection from the convex side? A parallel sheaf of light rays striking such a mirror is reflected away from the principal axis (except for the one ray that strikes right along the principal axis). Again, if the diverging reflected rays are continued backward (in imagination) through the mirror and beyond, they will converge to a focus.

The focus of a convex mirror, lying as it does behind the mirror, is a *virtual focus*, and its distance from the mirror is negative. For a convex mirror then, we must speak of $-f$ and, therefore, of $-1/f$. Again, since the reflected rays diverge, no real image will be formed in front of the mirror; only a virtual image (right-side up) behind the mirror. Therefore, we must speak of $-D_i$ and $-1/D_i$. For a convex mirror, Equation 2–2 becomes:

$$\frac{1}{D_o} - \frac{1}{D_i} = -\frac{1}{f}$$

(Equation 2–5)

or:

$$\frac{1}{D_o} = \frac{1}{D_i} - \frac{1}{f}$$

(Equation 2–6)

Since the object being reflected must always be in front of the mirror, D_o and, therefore, $1/D_o$ must be positive. It follows then that $1/D_i - 1/f$ must be positive, and for that to be true, $1/D_i$

Real and virtual images

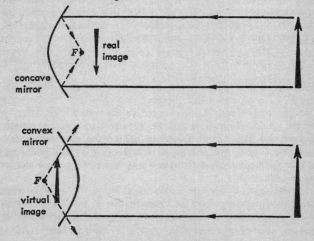

must be greater than $1/f$. But this leads us one step farther and tells us that D_i itself must be smaller than f. In other words, the apparent distance of all the virtual images reflected by a convex mirror must be less than that of the focus, however distant from the mirror the object being reflected is. For this reason, all objects reflected in a convex mirror seem compressed into a tiny space, and small convex mirrors at one corner of a large crowded room can give a panoramic view (albeit distorted) of the entire room.

The size of the image (S_i) is related to the size of the object being reflected (S_o), as the respective distances are related, regardless of whether those distances extend before or behind the mirror. In other words:

$$\frac{S_i}{S_o} = \frac{D_i}{D_o}$$

(Equation 2–7)

In a plane mirror, where the distance of the image from the mirror is equal to the distance from the mirror of the object being reflected, the sizes of the object and image are likewise equal. An image is neither diminished nor enlarged in a plane mirror. In a convex mirror, where all images must be closer to the mirror than the focus, however distant the objects being reflected are, all the images are small as well. The more distant the object being reflected, the closer and, therefore, the smaller the image.

In a concave mirror, however, when the object being reflected lies between the focus and the center of curvature, the image is beyond the center of curvature. In such a case, since the image is farther from the mirror than is the object being reflected, the image is larger than the object. The closer the object is brought to the focus, the larger the image appears. Of course, the larger the image is, the dimmer it is, for a given amount of light is spread out over a larger and larger volume.

Refraction

Light need not be reflected in order to deviate from straight-line motion. Light, in passing from one transparent medium into another, say from air into water, will generally not be reflected but will continue traveling forward and nevertheless may change direction.

This was undoubtedly first noticed by primitive men when a rod, placed in water in such a way that part remained in the air above, seemed bent at the point where it entered the water. If, however, it was withdrawn, it proved as straight and rigid as ever. Again, it is possible to place an object at the bottom of an empty cup and look at·the cup from such an angle that the object is just hidden by the rim. If water is now placed in the cup, the object at the bottom becomes visible though neither it nor the eye has moved. As long ago as the time of the ancient Greeks, it was realized that to explain this, one had to assume that light changed its direction of travel in passing from one transparent medium to another.

Imagine a flat slab of clear glass, perfectly transparent, and imagine a ray of light falling upon it along the normal line—that is, striking the glass at precisely right angles to its flat surface. The light, if one investigates the situation, is found to continue through the glass, its direction unchanged.

Suppose, though, that the light approaches the glass obliquely, forming the angle *i* with the normal. One might suspect that the light would simply continue moving through the glass, making the same angle *i* with the normal within the glass. This, however, is not what happens. The ray of light is bent at the point where air meets glass (the air-glass *interface*). Moreover, it is bent toward the normal in such a way that the new angle it makes with the normal inside the glass (*r*) is smaller than *i*, the angle of incidence.

This change in direction of a light ray passing from one transparent medium to another is called *refraction* (from Latin words meaning "to break back"). The angle *r* is, of course, the *angle of refraction*.

If the angle of incidence is made larger or smaller, the angle of refraction also becomes larger or smaller. For every value of *i*, however, where light passes from air into glass, *r* remains smaller.

The ancient physicists thought that the angle of refraction was directly proportional to the angle of incidence, and that therefore doubling *i* would always result in a doubling of *r*. This is nearly so where the angles involved are small, but as the angles grow larger, this early "law" fails.

Thus, suppose a light ray makes an angle of 30° to the normal as it strikes the air-glass interface and the angle of refraction that results after the light passes into the glass is 19.5°. If the angle of incidence is doubled and made 60°, the angle of refraction becomes 35.3°. The angle of refraction increases, but it does not quite double.

The correct relationship between *i* and *r* was worked out first in 1621 by the Dutch physicist Willebrord Snell (1591–1626). He did not publish his finding, and the French philosopher René

Descartes (1596–1650) discovered the law independently in 1637, publishing it in the form (rather simpler than Snell's) that we now use.

The Snell-Descartes *law of refraction* states that whenever light passes from one transparent medium into another, the ratio of the sine of the angle of incidence to the sine of the angle of refraction is constant.* The sine of angle *x* is usually abbreviated as sin *x*, so the Snell-Descartes law can be expressed:

$$\frac{\sin i}{\sin r} = n \qquad \text{(Equation 2–8)}$$

When a light ray passes (obliquely) from a vacuum into some transparent substance, the constant, *n*, is the *index of refraction* of that substance.

If light enters from a vacuum into a sample of gas at 0°C and 1 atmosphere pressure (these conditions of temperature and pressure are usually referred to as *standard temperature and pressure*, a phrase often abbreviated *STP*), there is only a very slight

* The sine of an angle can best be visualized as follows: Imagine the angle to be one of the acute angles of a right triangle. The sine of the angle is then equal to the ratio of the length of the side of the triangle opposite itself, to the length of the hypoteneuse of the triangle. Tables can be found in many texts and handbooks that give the values of the sines of the various angles. Thus, one can easily find that the sine of 10° 17′ is 0.17852, while the sine of 52° 48′ is 0.79653.

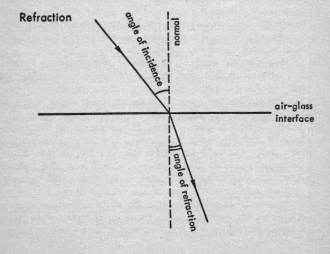

Refraction

angle of incidence

normal

air–glass interface

angle of refraction

refraction. This means that the angle of refraction is only very slightly smaller than the angle of incidence, and sin r is, in consequence, only very slightly smaller than sin i. Where this is true, we can see from Equation 2–8 that the value of n must be only very slightly greater than 1.

In fact, for hydrogen at STP, the index of refraction is 1.00013, and for air at STP it is 1.00029. There is very little error, therefore, in determining the index of refraction where light passes from air into some transparent substance rather than from vacuum into that transparent substance.

For liquids and solids, the situation is quite different. Water has an index of refraction of 1.33, while the index of refraction of glass varies from 1.5 to 2.0, depending on its exact chemical makeup. For an unusually high value, there is diamond, which has an index of refraction of 2.42. A ray of light entering diamond from air with an angle of incidence of 60°, passes into diamond with an angle of refraction of only 21.1°.

The greater the index of refraction of a material, the greater is its *optical density*. Thus, diamond is optically denser than glass, which is optically denser than water, which is optically denser than air. When a light ray travels from an optically less dense material into an optically more dense one, the direction of the light ray is bent toward the normal. This happens when light travels from air into water, for instance, or from water into diamond. A light ray traveling from an optically more dense material to an optically less dense one is bent away from the normal. One effect cancels the other. Thus if light passes from air into glass, striking at angle i and entering at angle r, and then passes from glass into air, striking at angle r, it will emerge at angle i.

Suppose, for instance, that a ray of light strikes a sheet of glass with an angle of incidence of 60°. The angle of refraction is 35.3°. After traveling through the thickness of the glass, the light ray reaches the other glass-air interface, which in the usual sheet of glass is precisely parallel to the first. As a result, any line which is normal to one interface is also normal to the other. At the second interface, the light is passing from glass into air, so it bends away from the normal. Since now it strikes at 35.3°, it emerges at 60°. The light having emerged from the glass sheet is now traveling in the same direction in which it had been traveling when it had entered; the refractive effect at one interface had been cancelled at the other, and the slight displacement of light rays that result goes unnoticed. (It is for this reason that looking obliquely through a window that is reasonably free of imperfections in the glass does not confuse us. Objects seen through a window are indeed in the direction they seem to be.)

Suppose we rearrange Equation 2–8 in order to solve for sin *r*. The result is:

$$\sin r = \frac{\sin i}{n} \qquad \text{(Equation 2–9)}$$

If the angle of incidence is 0°, then sin *i* is equal to 0, and sin *r* is equal to 0/*n*, or 0. The angle of incidence can be increased up to 90°, at which time the ray of light is perpendicular to the normal and just skims along parallel to the interface. If the angle of incidence has its maximum value of 90°, sin *i* is equal to 1, and the value of sin *r* is 1/*n*. In other words, as *i* goes through its extreme variation from 0° to 90°, sin *r* goes through an extreme variation from 0 to 1/*n*. In the case of water, where *n* equals 1.33, the extreme variation for sin *r* is from 0 to 0.75.

The angle that has a sine of 0 is 0°, and the angle that has a sine of 0.75 is (referring to a table of sines) 48.6°. Therefore as the angle of incidence for light passing from air into water varies from 0° to 90°, the angle of refraction varies from 0° to 48.6°. The angle of refraction cannot be higher than 48.6°, no matter what the angle of incidence.

But what if we reverse the situation and imagine light emerging from water into air? The relationship of the angles is reversed. Now the light is refracted away from the normal. As the light (in passing from water into air) forms an angle of incidence varying from 0° to 48.6°, the angle of refraction (formed by the light emerging into the air) varies from 0° to 90°.

Yet a skin diver under water with a flashlight may easily direct a beam of light so that it makes an angle to the normal (under water) of more than 48.6°. It should emerge at an angle of more than 90°, which means that it really does not emerge at all, since an angle of more than 90° to the normal will direct it under water again. The light, in other words, in passing from water to air will, if it strikes the interface at more than the *critical angle* of 48.6°, be reflected entirely. This is *total reflection*.

As you can see from Equation 2–9, the greater the index of refraction (*n*) of a substance, the smaller the critical angle. For ordinary glass the critical angle is about 42°, and for diamond, 24.5°. Light can be led through transparent plastic tubes around curves and corners if the rays from the light source, shining in at one end, always strike the plastic-air interface at angles greater than the critical angle for that plastic.

The index of refraction of air itself, while very small, can introduce noticeable effects where great thicknesses are involved. If a heavenly body is directly overhead, its light passes from the vacuum of space into the gas of our atmosphere with an incident angle of 0° and there is no refraction. An object that is not overhead has an angle of incidence greater than 0°, and its light is bent slightly toward the normal. Our eye, following the light backward without making allowance for any bending, sees the light source as somewhat higher in the sky than it actually is.

The lower in the sky a light source is, the greater the angle of incidence and the greater its difference from the angle of refraction. The greater, therefore, the discrepancy between its apparent and its real position. By the time objects at the horizon are involved, the eye sees an object higher than it really is by more than the width of the sun. Consequently, when the sun is actually just below the horizon, the refraction of the atmosphere allows us to see it as just above the horizon. Furthermore, the lowermost part of the sun, being lowest, undergoes the most refraction and is raised the more. As a result, the setting sun seems oval and flattened at the bottom.

Nor is the refractive curve of light as it enters our atmosphere from space a sharp one. The air is not uniformly dense but increases in density as one approaches earth's surface. Its index of refraction increases as its density does. Consequently, as light passes from space to our eye it bends more and more, following what amounts to a smooth curve (rather than the straight line we take so for granted).

The index of refraction of the air varies with temperature, too, and when a layer of air near the ground is heated, and overlaid with cooler air, light will curve in such a way as to make distant objects visible. The temperature conditions of the air may even cause objects on the ground to appear upside down in the air. The *mirages* that have resulted in this way (often in deserts where temperature differences between layers of air may be more extreme than elsewhere) have fooled victims all through history. In modern times such effects may make newspaper headlines, as when a person mistakes the headlights of a distant automobile reaching him through a long, gentle refractive curve, and reports "flying saucers" to be speeding their way through the sky.

CHAPTER 3

Lenses

Focus by Transmission

When the two edges of a piece of glass are not parallel, the normal to one edge will not be parallel to the normal to the other edge. Under such conditions, refraction at the far edge will not merely reverse the refraction at the near edge, and a ray of light passing through the glass will not emerge in the same direction it had on entrance. This is the case, for instance, when light passes through a triangle of glass, or *prism*.*

Imagine you are observing a ray of light touching the air-glass interface of such a prism, oriented apex-upward. If the ray of light meets the normal at an angle from below, it crosses into the glass above the normal but makes a smaller angle with it because the glass is optically more dense than the air. When the ray of light reaches the glass-air interface at the second side of the prism, it makes an angle with a new normal altogether, touching the interface above this normal. As it emerges into air, it must bend away from the new normal, because air is optically less dense than glass.

The result is that the ray of light bends twice in the same

* Any solid with parallel lateral edges and with a polygonal cross section when cut at an angle to those edges is a prism. Where the cross section is a triangle (*triangular prism*), we have the solid that is usually referred to simply as a prism, though it is actually only one example of an infinite class.

direction, first on entering the glass and then on leaving it. On leaving the glass it is traveling in a direction different from that in which it had entered. Light always passes through a prism in such a way that it bends away from the apex and toward the base.

Suppose you had two prisms set together, base to base, and a parallel beam of light is striking this double prism in a direction parallel to the mutual base line. The upper half of the beam, striking the upper prism, would be bent downward toward its base. The lower half of the beam, striking the lower prism, would be bent upward toward its base. The two halves of the beam of light, entering the double-prism parallel, would converge and cross on the other side.

The cross section of a double prism has interfaces consisting of two straight lines before and two straight lines behind, so its overall shape is that of a parallelogram (something like the "diamond" on the ace of diamonds in the deck of cards). In such a double prism, the normals to every point on the upper half are parallel because the interface is straight. Therefore all the rays in the light beam striking it make equal angles to the normal and are refracted through equal angles. The same is true for the lower half of the double prism, though there all the rays are bent upward rather than downward. The two half-beams emerge on the other side of the double prism as sheafs of parallel rays of light and cross each other over a broad front.

But what if the double-prism interfaces are smoothed out into a pair of spherical segments? The resulting figure would still be thin at the top and bottom and thickest in the middle, but now the normal to the surface would vary in direction at every point.

Prism

angle *a* greater than angle *b*
angle *d* greater than angle *c*

If the solid is held with its points in an up-down direction, then the normal would be horizontal at the center and would point more and more upward as one traveled toward the upper apex; it would point more and more downward as one traveled toward the lower apex.

Suppose now that a parallel sheaf of light rays strikes such a solid so that the ray striking the central thickest portion travels along the normal. It is not refracted but emerges from the other side unchanged in direction. Light rays striking a little above make a small angle with the upward-tipping normal and are refracted slightly downward. Light rays striking still higher make a somewhat larger angle with the further-tipping normal and are refracted downward more sharply, and so on. Below the center, the light rays are refracted upward more and more sharply as the distance from the center increases. The overall result is that the light rays converge on the other side of the lens, meeting at a focus.

A smoothed-out double prism of the type just described has the shape of a lentil seed and is therefore called a *lens* (from the Latin word for such a seed). By extension, any piece of glass or other transparent material with at least one curved surface is called a lens.

Both surfaces are convex in the particular type of lens that resembles a smoothed-out double prism. Such a lens is therefore a *biconvex lens*. This is the kind that resembles a lentil seed; it is the most familiar kind and, in fact, is what the average man will at once picture if asked to think of a lens.

It is not necessary for the two surfaces of a lens to be evenly curved. One surface might be less convex than the other, or it might even be flat. In the latter case, the lens is *plano-convex*. One of the surfaces can be concave for that matter (*concavo-convex*) so that a cross section of the lens looks something like a crescent moon. Such a lens may be called a *meniscus* (from a Latin word meaning "little moon"). Whatever the comparative shapes of the surfaces of the lens, light rays will be made to converge on passing through it, if the thickness of the lens is least at the edge and increases to maximum at the center. All such lenses can be lumped together as *convex lenses* or *converging lenses*.

The behavior of a convex lens neatly fits that of a convex mirror (see page 22). The light reflected from a convex mirror diverges, but if we imagine that the lines of the diverging rays are carried forward through the mirror, they will come to a focus on the other side; it is there that the virtual (upright) image is formed. In the case of a convex lens, light actually passes through and converges to a real focus where a real (inverted) image is formed. Because the image is real, light is concentrated and the ability of a lens to concentrate sunlight and start fires is well known.

The thicker the central bulge in a converging lens in relation to its diameter, the more sharply the rays of light are converged and the closer the focus is located to the lens itself—that is, the shorter the *focal length* (the distance from the focus to the center of the lens). A lens with a short focal length, which more drastically bends the light rays out of their original direction, is naturally considered a more powerful lens.

The strength of a lens is measured in *diopters* (from Greek words meaning "to see through"), which are obtained by taking the reciprocal of the focal length in meters. If the focal length is 1 meter, the lens has a power of $1/1$ or 1 diopter. A focal length of 50 centimeters, or 0.5 meters, implies a power of $1/0.5$, or 2 diopters. The larger the diopter value, the more powerful the lens.

A lens can be concave on both sides (a *biconcave lens*) so that it is thickest at the edges and thinnest in the center. It may be plane on one side (*plano-concave*) or even convex (*convexo-concave*). As long as it is thinnest in the center, it may be considered a *concave lens*. Since a parallel sheaf of light rays passing through any concave lens diverges after emerging on the other side, such lenses may also be called *diverging lenses*.

Here again, the properties of a concave lens and a concave mirror fit neatly. Light rays reflected from a concave mirror converge to a focus. If we imagine that the converging rays are carried through the mirror, they will diverge on the other side. In a concave lens, the light actually does pass through and diverge.

In the case of a concave lens, since the light passes through and diverges, it forms no image. However, the diverging light rays can be carried backward in imagination to form a virtual image on the forward side, where a concave mirror would have formed a real one.

The power of a diverging lens is arrived at in a manner similar to that in which the power of a converging lens is dealt with. However, in the case of a diverging lens, a virtual focus is involved, and the focal length therefore has a negative value. A diverging lens would be said to have a power of -2 diopters, for instance.

Spectacles

There is a lens-shaped object within the human eye, just behind the pupil, which is called the *crystalline lens* (not because it contains crystals, but because, in the older sense of the word "crystalline," it is transparent). It is a biconvex lens, and therefore a converging lens, about a third of an inch in diameter. The foremost portion of the eye, the transparent *cornea*, is also a converging lens, with twice the converging power of the crystalline lens itself.

The cornea and crystalline lenses converge the light rays to a focus upon the light-sensitive inner coating (*retina*) of the rear of the eyeball. An inverted image forms on the retina, and the pattern of light and dark is imprinted there. Each light-sensitive cell in the retina's center (where the image of what we are looking at is formed) is connected to an individual nerve fiber, so the pattern is carried without loss of detail to the brain. The brain makes allowance for the inversion of the image, and we see right-side-up.

The image formed by a converging lens cannot, however, always be counted upon to fall upon the focus (which, strictly speaking, is the point at which a sheaf of parallel rays of light are made to converge). Where the light source is far away, the rays are indeed parallel or virtually so, and all is well. As the light source is brought nearer to the lens, however, the light rays are more and more perceptibly divergent, and they then converge beyond the focus—that is, at a distance greater than the focal length.

The relationship between the distances of the object serving as the light source (D_o), of the image (D_i), and of the focus (f) can be expressed by means of Equation 2–2 (see p. 18). In the previous chapter, this equation was used in connection with mirrors, but it will serve for lenses, too. In fact, it is so commonly used for lenses rather than for mirrors that it is usually called the *lens formula*. (In both lenses and mirrors a virtual focus yields a negative value for f and $1/f$, and a virtual image, a negative value for D_i and $1/D_i$. On the other hand, D_o and $1/D_o$ are always positive.)

Let us rearrange the lens formula and write it as follows:

$$\frac{1}{D_i} = \frac{1}{f} - \frac{1}{D_o}$$

(Equation 3–1)

If the object is at an infinite distance, $1/D_o = 0$, and $1/D_i = 1/f$, which means that $D_i = f$. The image, therefore, is formed at the focus. But let us suppose that the focal length of the cornea-lens combination of the eye is about 1.65 centimeters (which it is) and that we are looking at an object 50 meters (or 5000 centimeters) away. In that case, $1/D_i = 1/1.65 - 1/5000$, and $D_i = 1.6502$. The image forms 0.0002 centimeters beyond the focus, a discrepancy that is small enough to be unnoticeable. Thus, a distance of 50 meters is infinite as far as the eye is concerned.

But what if the object were 30 centimeters away—reading distance? Then $1/D_i = 1/1.65 - 1/30$, and $D_i = 1.68$. The image would form about 0.03 centimeters behind the focus, and on the scale of the eye that would be a serious discrepancy. The light would reach the retina (at the focal length) before the light rays had focused. The image would not yet be sharp, and vision would be fuzzy.

To prevent this, the crystalline lens changes shape through the action of a small muscle. It is made to thicken and become a more powerful light-converger. The focal length shortens. The image, still forming beyond the new and now shorter focal length, forms on the retina. This process is called *accommodation*.

As an object comes nearer and nearer the eye, the crystalline lens must bulge more and more to refract the light sufficiently to form the image on the retina. Eventually it can do no more, and the distance at which accommodation reaches its limit is the *near-point*. Objects closer to the eye than the near-point will seem fuzzy because their image cannot be made to form on the retina.

The ability to accommodate declines with age, and the near-point then recedes. A young child with normal vision may be able to focus on objects as close to the eye as 10 centimeters; a young adult on objects 25 centimeters away; while an old man may not be able to see clearly anything closer than 40 centimeters. In other words, as one grows older one starts holding the telephone book farther away. This recession with age of the near-point is called *presbyopia* (from Greek words meaning "old man's vision").

It may happen that a person's eyeball is deeper than the focal length of the cornea-lens combination. In such a case, the images of objects at a distance form at the focus, which is well in front of the too-deep retina. By the time the light rays reach the retina, they have diverged a bit and vision is fuzzy. As objects come closer, the image is formed at distances greater than the focal length, and these eventually do fall on the retina. Such people can clearly see near objects but not distant ones; they are *nearsighted*. More formally, this is called *myopia*, for reasons to be explained on page 39.

The opposite condition results when an eyeball is too shallow. The focal length is greater than the depth of the eyeball, and the light rays reaching the retina from objects at a great distance have not yet quite converged. The crystalline lens accommodates and bends the light more powerfully so that distant objects can be seen clearly after all. As an object approaches more closely, however, the power of lens accommodation quickly reaches its limit, and near objects can only be seen fuzzily. For such a person, the near-point is abnormally far away, and although he can see distant objects with normal clarity, he cannot see near objects clearly. He is *farsighted* and suffers from *hyperopia* ("vision beyond").

It is easy to produce a new overall focal length by placing one lens just in front of another. One need, then, only add the diopters of the two lenses to find the total refracting power of the two together, and therefore the focal length of the two together.

Imagine a lens with a refracting power of 50 diopters. Its focal length would be 1/50 of a meter, or 2 centimeters. If a second converging lens of 10 diopters were placed in front of it, the refracting power of the lens combination would be 60 diopters, and the new focal length would be 1/60 of a meter, or 1 2/3 centimeters. On the other hand, a diverging lens with a refracting power of —10 diopters would increase the focal length, for the two lenses together would now be 40 diopters and the focal length would be 1/40 meter of 2 1/2 centimeters.

This can be done for the eye in particular and was done as early as the thirteenth-century by such men as the English scholar Roger Bacon (1214?–1294). The results are the familiar *eyeglasses* or *spectacles*, and these represent the one great practical application of lenses that was introduced during the Middle Ages.

The power of the cornea-lens combination of the eye is about 60 diopters, and the lenses used in spectacles have powers ranging from —5 to +5 diopters. For farsighted people with too-shallow eyeballs, the diopters must be increased so that focal length is decreased. To increase the diopters, a lens with positive diopters (that is, a converging lens) must be placed before the eye. The reverse is the case for nearsighted individuals. Here the eyeball is too deep, and so the focal length of the eye must be lengthened by reducing the diopters. A lens with negative diopters (that is, a diverging lens) must be placed before the eye.

For both farsighted and nearsighted individuals, the spectacle lens is usually a meniscus. For the former, however, the meniscus is thickest at the center; for the latter it is thinnest at the center.

As old age comes on, the additional complication of presbyopia may make it necessary to apply two different corrections, one for near vision and one for far vision. One solution is to have two different types of spectacles and alternate them as needed. In his old age, it occurred to the American scholar Benjamin Franklin (1706–1790), when he grew weary of constantly switching glasses, that two lenses of different diopters, and therefore of different focal lengths, could be combined in the same frame. The upper portion might be occupied by a lens correcting for far vision, the lower by one correcting for near vision. Such *bifocals* (and occasionally even *trifocals*) are now routinely produced.

For a lens to focus reasonably well, its curvature must be the same in all directions. In this way, the rays that strike toward the top, bottom, and side of the lens are all equally converged toward the center, and all meet at a true focus.

Suppose the lens curves less sharply from left to right than from top to bottom. The light rays at left and right would then not come to a focus at a point where the light rays from top and bottom would. At that point, instead of a dot of light, there would be a horizontal line of light. If one moves farther back, to a spot where the laggard rays from right and left have finally focused, the rays from top and bottom have passed beyond focus and are diverging again. Now there is a vertical line of light. At no point is there an actual dot of light. This situation is common with respect to the eyeball, and the condition is called *astigmatism** (from Greek words meaning "no point"). This, too, can be corrected by using spectacles that have lenses with uneven curvatures that balance the uneven curvature of the eye, bending the light more in those directions where the eye itself bends it less.

The usual lenses are ground to the shape of segments of spheres, since the spherical shape is the easiest to produce. Such a shape, even if perfectly even in curvature in all directions, still does not converge all the rays of light to an exact point, any more than a spherical mirror reflects all the rays to an exact point. There is spherical aberration (see page 17) here as well as in the case of mirrors.

---

* The term "astigmatism" as applied to lenses is not quite the same in meaning as when it is applied to the eye. In lenses, it is produced when the source of light is not on the principal axis of the lens. Light in that case strikes the lens obliquely and is not brought to perfect focus but to a line of light.

The extent of this aberration increases with the relative thickness of the lens and with distance from the center of the lens. For this reason, the lens formula (Equation 3–1) holds well only for thin lenses. Near the center of the lens, the spherical aberration is quite small and can usually be ignored. The human eye is fitted with an iris that can alter the size of the pupil. In bright light, the size of the pupil is reduced to a diameter of 1.5 millimeters. The light that enters is still sufficient for all purposes, and spherical aberration is reduced to almost nothing. In bright light, therefore, one sees quite clearly. In dim light, of course, it is necessary to allow as much light to enter the eye as possible, so the pupil expands to a diameter of as much as eight or nine millimeters. More of the lens is used, however, and spherical aberration increases. In dim light, therefore, there is increased fuzziness of sight.

There are other types of aberration (including "chromatic aberration," see page 55), but the usual way of correcting such aberrations in elaborate optical instruments is to make use of two lenses in combination (or a mirror and a lens) so that the aberration of one will just cancel the aberration of the other. By a clever device of this sort, in 1930 a Russian-German optician, Bernard Schmidt (1879–1935) invented an instrument that could without distortion take photographs over wide sections of the sky because every portion of its mirror had had its aberrations canceled out by an irregularly shaped lens called a "corrector plate." (Such an instrument is called a Schmidt camera or a Schmidt telescope.)

Cameras

Images can be formed outside the eye, of course, as well as inside. Consider a single point in space and an object, some distance away, from which light is either being emitted or reflected. From every part of the object a light ray can be drawn to the point and beyond. A ray starting from the right would cross over to the left once it had passed the point, and vice versa. A ray starting from the top would cross over to the bottom once it had passed the point, and vice versa.

Suppose the rays of light, having passed the point, are allowed to fall upon a dark surface. Light rays from a brightly emanating (or reflecting) portion of the light source would yield bright illumination; light rays originating from a dimly lit portion would yield dim illumination. The result would be a real and inverted image of the light source.

Actually, under ordinary conditions we cannot consider a single point in space, since there are also a vast number of neighboring points through which rays from every portion of the light source can be drawn. There are, therefore, a vast number of inverted images that will appear on the surface, all overlapping, and the image is blurred out into a general illumination; in effect, no images are formed.

But suppose one uses a closed box with a hole on the side facing the light source, and suppose one imagines the hole made smaller and smaller. As the hole is made smaller, the number of overlapping images is continually being reduced. Eventually an image with fuzzy outlines can be made out on the surface opposite the hole, and if the hole is made quite small, the image will be sharp. The image will remain sharp no matter what the distance between the hole and the surface on which it falls, for there is no question of focusing since the image is formed of straight-line rays of light that are unrefracted. The farther the surface from the hole the larger the image, since the rays continue to diverge with increasing distance from the hole. However, because the same amount of light must be spread over a larger and larger area, the image grows dimmer as it grows larger.

On a large scale, this can be done in a dark room with the windows thickly curtained except for one small hole. On the opposite wall an image of whatever is outside the hole will appear —a landscape, a person, a building—upside down, of course.

The sun shining through such a hole will form a circle that is actually the image of the sun, and not of the hole. If the hole were triangular in shape but not too small, there would be a triangular spot of light on the wall, but this triangle would be made up of circles, each one of which would be a separate image of the sun. As the hole grows smaller, so does the triangle, until it is smaller than an individual circular image of the sun. At that point, the image will appear a circle despite the triangularity of the hole.

The leaves of a tree form a series of small (though shifting) openings through which sunlight streams. The dappled light on the ground then shows itself as small superimposed circles, rather than reproducing the actual irregular spaces between the leaves. During a solar eclipse, the sun is no longer round but is bitten into and, eventually, shows a crescent shape. When this takes place, the superimposed circles of light under the tree becomes superimposed crescents. The effect is quite startling.

Image-formation in dark rooms began in early modern times,

and such Italian scholars as Giambattista della Porta (1538?–1615) and Leonardo da Vinci (1452–1519) made use of it. The device is called a *camera obscura*, which is a Latin phrase meaning "dark room." Eventually other devices for producing images within a darkened interior were used, and the first part of the phrase, "camera," came to be applied to all such image-forming devices. The original camera obscura, with its very small opening, is now commonly called a *pinhole camera*.

The chief difficulty with a pinhole camera is that to increase the sharpness of the image one must keep the hole as small as possible. This means that the total amount of light passing through the hole is small, and the image is dim. To widen the opening and allow more light to enter, and yet avoid the superimpositions that would immediately destroy the image, one must insert a converging lens in the opening. This will concentrate the light from a large area into a focus, increasing the brightness of the image many times over without loss of sharpness. In 1599, della Porta described such a device and invented the camera as we now know it.

Once a camera is outfitted with a lens, the image will no longer form sharply at any distance, but only at the point where the light rays converge. For cameras of fixed dimensions, sharp images may be formed only of relatively distant objects, if the back of the camera is at the focal length. For relatively close objects, the light rays converge at a point beyond the focal length (see page 34) and the lens must be brought forward by means of an accordionlike extension (in old-fashioned cameras) or by means of a screw attachment (in newer ones). This increases the distance between the lens and the back of the camera, and is the mechanical analog of the eye's power of accommodation.

In an effort to make out objects in the middle distance, people who are nearsighted quickly learn that if they squint their eyes they can see more clearly. This is because the eye is then made to approach more closely to the pinhole camera arrangement, and a clear image depends less on the depth of the eyeball. (Hence "myopia" is the term used for nearsightedness, for this comes from a Greek phrase that means "shut-vision" with reference to the continual squinting.) Of course, the difficulty is that less light then enters the eye, so sharper focus is attained at the expense of brightness. Furthermore, the muscles of the eyelids tire of the perpetual task of keeping them somewhat but not altogether closed; the result is a headache. (Actually, it is "eye-muscle strain" and not "eyestrain" that causes the discomfort.)

The lensed camera came of age when methods were discovered for making a permanent record of the image. The image is formed upon a surface containing chemicals that are affected by light.* A number of men contributed to this, including the French physicist Joseph Nicéphore Niepce (1765–1833), the French artist Louis Jacques Mandé Daguerre (1789–1851), and the English inventor William Henry Fox Talbot (1800–1877). By the mid-nineteenth century, the camera as producer and preserver of images was a practical device, and *photography* ("writing by light") became of infinite use in every phase of scientific work.

To get bright images, as much light as possible must be squeezed together. This requires a lens of large diameter and short focus. The larger the diameter, the more light is gathered together and converged into the image. The reason for the short focus depends on the fact (already discussed in connection with mirrors on page 23 and a point applicable in the case of lenses as well) that the closer the image to the lens, the smaller it is. The smaller the image into which a given quantity of light is focused, the brighter it is. To measure the brightness of the image that a lens can produce, we must therefore consider both factors and take the ratio of the focal length (f) to the diameter (D). This ratio, f/D, is called the *f-number*. As one decreases f or increases D (or both), the f-number decreases. The lower the f-number, the brighter the image.

The image, as originally formed on chemically-coated film, is dark in spots where intense illumination has struck (for the effect of light is to produce black particles of metallic silver) and light where little illumination has struck. The image therefore appears in reverse—light where we see dark and dark where we see light. This is a *negative*. If light is projected through such a negative onto a paper coated with light-sensitive chemicals, a negative of the negative is obtained. The reversal is reversed, and the original light-dark arrangement is obtained. This *positive* is the final picture.

The positive may be printed on transparent film. In that case, a small but intense light source may be focused upon it by a lens and mirror combination, and the image projected forward onto a screen. The rays diverge after leaving the *projector*, and the image on the screen can be greatly enlarged, as compared with the original positive. This can be used for home showing of photo-

* The details of the process are more appropriate for a book on chemistry, and will not be considered here.

graphs, and has been used, far more importantly, as a means of mass entertainment.

The possibility for this arises from the fact that when the cells of the retina react to a particular pattern of light and dark, it takes them a perceptible fraction of a second to recover and be ready for another pattern. If, in a dark room, you wave a long splint, smoldering at the far end, you will not see a distinct point of light changing position, but a connected curve of light out of which you can form circles and ovals.

Imagine, then, a series of photographs taken very rapidly of moving objects. Each photograph would show the objects in slightly different positions. In 1889, the American inventor Thomas Alva Edison (1847–1931) took such photographs on a continuous strip of film with perforations along the side. Such perforations could be threaded onto a sprocket wheel, which, when turning, would pull the film along at a constant velocity. If a projector light could be made to flash on and off rapidly, it would flash onto the screen a quick image of each passing picture. The eye would then see one picture after another, each just slightly different from the one before. Because the eye would experience its lag period in reaction, it would still be seeing one picture when the next appeared on the screen. In this way, an illusion of continual change, or motion, is produced. Thus, *motion pictures* were introduced.

Magnification

Anyone who has handled a converging lens knows that objects viewed through it appear larger. It is very likely that this was known in ancient times, since a round glass bowl filled with water would produce such an effect.

To understand this, we must realize that we do not sense the actual size of an object directly, but merely judge that size from a variety of indirect sensations, including the angle made by light reaching the eye from extreme ends of the object.

Suppose, for instance, that a rod 4 centimeters long is held horizontally 25 centimeters in front of the eyes. The light reaching the eye from the ends of the rod makes a total angle of about 9.14°. In other words, if we looked directly at one end of the rod, then turned to look directly at the other end, we would have turned through an angle of 9.14°. This is the *visual angle*, or the *angular diameter* of an object.

If the rod were only 2 centimeters long, the visual angle would be 4.58°; if it were 8 centimeters long, it would be 18.18°. The visual angle is not exactly proportional to the size, but for small values it is almost exactly proportional. We learn this proportionality through experience and automatically estimate relative size by the value of the visual angle.

However, the angular size of any object is also a function of its distance. Consider the eight-centimeter rod that at 25 centimeters would exhibit a visual angle of 18.18°. At 50 centimeters its visual angle would be 9.14°; at 100 centimeters, 4.58°. In other words, as we also know from experience, an object looks smaller and smaller as it recedes from the eye. A large object far distant from the eye would look smaller than a small object close to the eye. Thus, an eight-centimeter rod 100 centimeters from the eye would produce a smaller visual angle than a four-centimeter rod 25 centimeters from the eye, and the former would therefore appear smaller in size.

It is not likely that we would be fooled by this. We learn at an early age to take distance, as well as visual angle, into account in assessing the real size of an object. In looking first at the distant eight-centimeter rod then at the closeby four-centimeter rod, we must alter the accommodation of the crystalline lens, and we must also alter the amount by which our eyes have to converge in order for both to focus on the same object (the closer the object, the greater the degree of convergence). We may not be specifically aware that our lenses are accommodating and our eyes converging; however, we have learned to interpret the sensations properly, and we can tell that the four-centimeter rod is closer. Making allowance for that as well as for the visual angle, we can usually tell without trouble that the rod that looks smaller is actually larger. We even convince ourselves that it *looks* larger.

Alterations in the accommodation of the lens and the convergence of the eyes are of use only for relatively nearby objects. For distant objects, we judge distance by comparison with neighboring objects whose real size we happen to know. Thus a distant sequoia tree may not look unusually large to us until we happen to notice a tiny man at its foot. We then realize how distant it must be, and its real size is made apparent. It begins to look large.

If there are no neighboring objects of known size with which to compare a distant object, we have only the visual angle, and that by itself tells us nothing. For instance, the moon, high in the sky, presents a visual angle of roughly 0.5°. If we try to judge the real diameter of the moon from this, we are lost. We might decide that the moon looked "about a foot across." However, an object a foot across will produce a visual angle of 0.5° if it is not quite sixty feet away. This is certainly a gross underestimate of the actual distance of the moon, yet many people seem to assume, unconsciously, that that is the distance.

When the moon is near the horizon, it is seen beyond the houses and trees, and we know at once that it must be more than sixty feet away. It might be, let us say, a mile away. To produce a visual angle of 0.5° from a distance of a mile, the moon would have to be 88 feet across. This (unconscious) alteration in our estimate of the moon's distance also alters our (unconscious) estimate of its real size. The moon, as all of us have noticed, seems much larger at the horizon than when it is high in the sky.* This optical illusion has puzzled men ever since the time of the Greeks, but the present opinion of men who have studied the problem is that it is entirely a matter of false judgment of distance.

A converging lens offers us a method for altering the visual angle without altering the actual distance of an object. Consider light rays traveling from an object to the eye and making a certain visual angle. If, on the way, they pass through a converging lens, the light rays are converged and make a larger visual angle. The eye cannot sense that light rays have been converged en route; it judges the light rays as though they came in straight lines from an object larger than the real object. Only by sensing the object as enlarged, can the eye account for the unusually large visual angle. Another way of putting it is that the eye sees not the object

* Actual measurement of the moon's apparent diameter shows that at the horizon it is actually a tiny bit smaller than at zenith, for at the horizon the radius of the earth must be added to its distance. It is therefore 2 percent farther from our eyes at the horizon, and its visual angle is 2 percent smaller.

Magnification

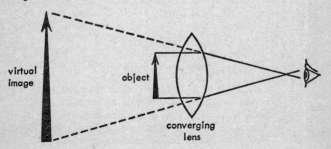

virtual image

object

converging lens

but an enlarged virtual image (hence right-side-up) of the object on the other side of the converging lens. The ratio of the size of the image to the size of the object itself is the *magnification* of the lens.

The magnification can be expressed in terms of the focal length (f) of the lens, provided we turn once more to the lens equation (Equation 2–2 on page 24, or 3–1 on page 33). Since the image is a virtual image, its distance (D_i) receives a negative sign, while the distance of the object itself (D_o) remains positive, as always. The equation can be written then:

$$\frac{1}{D_o} - \frac{1}{D_i} = \frac{1}{f}$$ (Equation 3–2)

The magnification, as I have said, is the ratio of the size of the image to the size of the object, but this size ratio can be judged in two ways. It can be interpreted as the ratio of the visual angles, if both object and image are at the same distance; or as the ratio of the distances, if both object and image produce the same visual angle. Let us take the latter interpretation and solve Equation 3–2 for the ratio of the distance of the image to that of the object (D_i/D_o). It turns out that:

$$\frac{D_i}{D_o} = \frac{f}{f - D_o} = m$$ (Equation 3–3)

where m is the magnification.

If the lens is held in contact with the object, which may be a printed page, for instance, D_o is virtually zero and $f - D_o = f$. The magnification m is then equal to f/f, or 1, and the print is not magnified. If the lens is lifted, D_o increases, which means that $f - D_o$ must decrease and, as you can see from Equation 3–3, m must, in consequence, increase. The print seems larger and larger as the lens is lifted. When the distance of the lens from the printed page is equal to the focal length, $f - D_o$ becomes equal to $f - f$, or 0. Magnification is then $f/0$ and becomes infinite. However, no lens is perfect, and if the object is magnified infinitely, so are all the imperfections. As a result, all that can be seen is a blur. Maximum practical magnification comes when the distance of the object is just a little short of the focal length.

If the object is at a distance greater than the focal length, $f - D_o$ becomes negative and therefore m becomes negative. As D_o continues to increase in size, m remains negative, but its absolute value (its value with the negative sign disregarded) becomes smaller. This means that the image becomes inverted and decreases in size again as the object distance becomes greater than the focal length and continues to increase.

It also follows from Equation 3–3 that, for a given distance of the object (D_o), the magnification increases with decrease in the focal length of the lens (provided the focal distance remains greater than the distance of the object). To see this, let us suppose that $D_o = 1$ and that f takes up successive values of 5, 4, 3 and 2. Since the magnification (m) is equal to $f/(f - D_o)$, it equals, successively, 5/4, 4/3, 3/2 and 2/1; or 1.2, 1.33, 1.5 and 2.0. This is another reason for considering a converging lens to grow more powerful as its focal length decreases (see page 40). After all, its magnifying power increases as its focal length decreases.

All this is reversed for diverging lenses. Here the rays of light converging on their trip to the eye from opposite ends of an object are diverged somewhat by the lens and made to reach the eye at a smaller visual angle. For that reason, objects seem smaller when viewed through a diverging lens.

In this way, you can quickly tell whether a person is nearsighted or farsighted by an extremely simple test with his glasses. A nearsighted man must wear diverging lenses (see page 35), so print looks smaller if those lenses are held a few inches above the printed page. A farsighted man must wear converging lenses, and those will make the print appear larger.

Microscopes and Telescopes

The cells of the retina either "fire" as light strikes them, or do not fire as light does not strike them. As a result, the image that is produced upon them is, so to speak, a combination of light and dark spots. This resembles the appearance of a newspaper halftone reproduction, though the "spots" on the retina are much finer than those on a newspaper photograph.

When an object is considerably larger than the spots that make it up, the object is seen clearly. If it is not much larger, it is seen fuzzily. Thus, if you look at a newspaper photograph with the unaided eye, you will seem to see a clearly delineated face. If you look at it under a magnifying lens, the portion you see in the lens will not be much larger than the magnified dots, and things will not be clear at all. You will not make out "detail."

In the same way, there is a limit to the amount of detail you can see in any object with the unaided eye. If you try to make out finer and finer details within the object, those details begin to be no larger (in the image on your retina) than the dots making up the image. The retinal image becomes too coarse for the purpose.

Light from two dots separated by an angular distance of less than a certain crucial amount activates the same retinal cell or possibly adjacent ones. The two dots are then seen as only a single dot. It is only when light from two dots activates two retinal cells separated by at least one unactivated cell that the two dots can actually be seen as two dots. At 25 centimeters (the usual distance for most comfortable seeing) two dots must be separated by at least 0.01 centimeters to be seen as two dots; the minimum visual angle required is therefore something like 0.006°.

The *resolving power* of the human eye (its ability to see two closely-spaced dots as two dots and, in general, its ability to make out fine detail) is actually very good and is much better than that of the eyes of other species of animals. Nevertheless, beyond the resolving power of the human eye there is a world of detail that would be lost to our knowledge forever, were it not for lenses.

Suppose two dots, separated by a visual angle of 0.001°, were placed under a lens with a magnification of 6. The visual angle formed by those two dots would be increased to 0.006°, and they could be seen as two dots. Without the lens, they could be seen as only one dot. In general, an enlarging lens not only makes the object larger in appearance, it makes more detail visible to the eye.

To take advantage of this, one must use good lenses, that have smoothly ground surfaces and are free of bubbles and imperfections. A lens that is not well constructed will not keep the refracted light rays in good order, and the image, though enlarged, will be fuzzy. Fine detail will be blurred out and lost.

It was not until the seventeenth century that lenses accurate enough to keep at least some of the fine detail were formed. A Dutch merchant, Anton van Leeuwenhoek (1632–1723), used small pieces of glass (it is easier to have a flawless small piece of glass than a flawless large one) and polished them so accurately and lovingly that he could get magnifications of more than 200 without loss of detail. With the use of such lenses, he was able to see blood capillaries, red blood corpuscles, and spermatozoa. Most important of all he could study the detail of independently living animals (protozoa) too small to be made out with the naked eye.

Such strongly magnifying lenses are *microscopes* (from Greek words meaning "to see the small"). A microscope made out of a single lens, as Leeuwenhoek's were, is a *simple microscope*.

There is a limit to the magnifying power of a single lens, however well-ground it may be. To increase the magnifying power, one must decrease the focal length, and Leeuwenhoek was already using minute focal lengths in his tiny lenses. It would be impractical to expect much further improvement in this respect.

However, suppose the light from an object is allowed to pass through a converging lens and form a real image on the other side. As in the case of concave mirrors (see page 23), this real image may be much larger than the object itself, if the object is quite near the focus. (The image would then be much dimmer because the same amount of light would be spread over a greater area. For this reason, the light illuminating the object must be quite intense in the first place, in order to remain bright enough despite this dimming effect.)

Since the image is a real image, it can be treated optically as though it were the object itself. A second converging lens can be used that will further magnify the already-magnified image. By the use of two or more lenses in this way, we can easily obtain a final magnification that will be greater than the best one can do with a single lens. Microscopes using more than one lens are called *compound microscopes*.

The first compound microscopes are supposed to have been built a century before Leeuwenhoek by a Dutch spectacle maker, Zacharias Janssen, in 1590. Because of the imperfect lenses used, it took quite a while for these to be anything more than playthings. By the end of Leeuwenhoek's life, however, compound microscopes were beginning to surpass anything his simple lenses could do.

The *telescope* (from Greek words meaning "to see the distant") also makes use of lenses. Light from an object such as the moon, let us say, is passed through a converging lens and allowed to form a real image on the other side. This image is then magnified by another lens. The magnified image is larger and shows more detail than the moon itself does when viewed by the naked eye.

A telescope can be used on terrestrial objects, too. Here, since the real image formed through the converging lens is inverted, and it would be disconcerting to see a distant prospect with the ground above and the sky below, two lenses are used to form the image, the second lens inverting the inverted image and turning it right-side-up again. This new right-side-up image can then be magnified, and we have a *field glass* for use on landscapes. Small field glasses, designed in pairs to be looked through with both eyes at once, are *opera glasses*.

Astronomical telescopes do not make use of the additional lens, since each lens introduces imperfections and problems, and the fewer lenses the better. An upside-down star or moon does not disconcert an astronomer, and he is willing to let the image remain that way.

The telescope is supposed to have been invented by an apprentice-boy in the shop of the Dutch spectacle maker Hans Lippershey in about 1608.* The next year, the Italian scientist Galileo Galilei (1564–1642), concerning whom I had occasion to speak at length in the first volume of this book, hearing rumors of this new device, experimented with lenses until he had built a telescope. His instrument was exceedingly poor in comparison with modern ones; it only enlarged about thirtyfold. However, in turning it on the sky, he opened virgin territory, and wherever he looked, he saw what no man had ever seen before.

The greater detail visible on the image of the moon made it possible for him to see lunar mountains and craters. He saw spots on the sun and enlarged both Jupiter and Venus into actual globes. He could see that Venus showed phases like the moon (as was required by the Copernican theory) and that Jupiter was circled by four satellites.

The lens of a telescope also serves as a light-collector. All the light that falls upon the lens is concentrated into the image. If the lens is larger than the pupil of the eye (and in a telescope it is bound to be), more light will be concentrated into the image within the telescope than is concentrated into the image within the eye. A star that is too dim to be made out by the unaided eye becomes bright enough therefore to be seen in a telescope. When Galileo turned his telescope on the starry sky, he found a multiplicity of stars that were plainly visible with the telescope and that vanished when he took the instrument from his eye.

Naturally, the larger the lens, the more light it can collect, and the dimmer the stars it can make out. The Yerkes telescope of today (a distant descendant of the first Galilean telescope) has a collecting lens 40 inches in diameter, as compared with the pupil's diameter of no more than 1/3 of an inch. The ratio of the diameters is thus 120 to 1. The light collected depends on the area of the lens, and this is proportional to the square of the diameter. The light-collecting power of the Yerkes telescope is therefore 14,400 times as great as that of the human eye, and stars correspondingly dimmer can be made out by it.

Furthermore, if the light from the telescope is focused on photographic film, rather than on the retina of the eye, there is a further advantage. Light striking the film has a cumulative effect

---

* The boy was playing with lenses when he should have been working, and you can draw your own moral from that.

(which it does not have on the eye). A star too dim to be seen, even through the telescope, will slowly affect the chemicals on the film and after an appropriate time exposure, can be photographed even if it cannot be seen.

In theory, lenses can be made larger and larger, and the universe probed more and more deeply. However, practical considerations interfere. The larger the lens, the more difficult and tedious it is to grind it exactly smooth and the more difficult it is to keep it from bending out of shape under its own weight (since it can only be supported about the rim). In addition, the larger the lens, the thicker it must be, and since no lens is perfectly transparent, the thicker it is, the more light it absorbs. After a certain point, it is impractical to build larger lenses. The telescope at the Yerkes Observatory in Wisconsin has a 40-inch lens and is the largest telescope of its sort in the world. It was built in 1897 and nothing larger has been built since. Nor is any likely to be built.

4

Color

Spectra

So far, I have spoken of light as though all light were the same except that one beam might differ from another in brightness. Actually, there is another distinguishing characteristic, familiar to us all, and that is *color*. We know that there is such a thing as red light, blue light, green light, and so on through a very large number of tints and shades.

The tendency in early times was to consider the white light of the sun as the simplest form of light—as "pure" light. (Indeed, white is still the symbol of purity, and the young bride walks to the altar in a white wedding gown for that reason.) Color, it was felt, was the result of adding impurity to the light. If it traveled through red glass, or were reflected from a blue surface, it would pick up redness or blueness and gain a property it could not have of itself.

From that point of view it would be very puzzling if one were to find the pure white light of the sun displaying colors without the intervention of colored matter at any time. The one such phenomenon known to men of all ages is the *rainbow*, the arc of varicolored light that sometimes appears in the sky, when the sun emerges after a rainshower. The rainbow was startling enough to attract a host of mythological explanations; a common one was that it was a bridge connecting heaven and earth. The first motion toward a rationalistic explanation was that of the Roman philosopher Lucius Annaeus Seneca (4 B.C.?–65 A.D.), who pointed out that the rainbow was rather similar to the play of colors often seen at the edge of a piece of glass.

By the seventeenth century, physicists began to suspect that the rainbow, as well as the colors at the edge of glass, were somehow produced by the refraction of light. The French mathematician René Descartes worked out a detailed mathematical treatment of the refraction and total reflection of light by spheres of water. In this way he could account nicely for the position of the rainbow with relation to the sun, thanks to the refraction of sunlight by tiny droplets of water remaining suspended in air after the rain, but he could not account for the color.

It was left for the English scientist Isaac Newton—whose work takes up so much of Volume I of this book—to make the crucial advance. In 1666, he allowed a shaft of sunlight to enter a darkened room and to fall on a prism. The beam of light, refracted through the prism, was allowed to strike a white surface. There it appeared, not as a spot of white sunlight, but as an extended band of colors that faded into one another in the same order (red, orange, yellow, green, blue, violet) they do in a rainbow. It was a colored image and it received the name of *spectrum*, from the Latin word for "image."

If the light of the spectrum was formed on a surface with a hole in it, a hole so arranged that only one of the colors could pass through, and if that one beam of colored light was allowed to pass through a second prism, the color would be somewhat spread out but no new colors would appear.

Newton's contribution was not that he produced these colors, for that had been done before, but that he suggested a new ex-

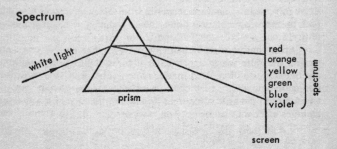

Spectrum

white light

prism

red
orange
yellow
green
blue
violet

spectrum

screen

planation for them. The only ingredients that produced the spectrum were the pure white light of sunlight and the pure colorless glass of the prism. Newton therefore stated that despite the long-settled opinion of mankind white light was not pure, but a complex mixture of all the colors of the rainbow. It appeared white only because the combination so stimulated the retina as to be interpreted by the brain as "white."

As a point in favor of the validity of this view, there is Newton's reversal of spectrum formation. Newton allowed the light of the colored spectrum to fall upon a second prism held upside down with respect to the first. The light was refracted in the opposite direction and the situation was reversed. Where previously a round beam of white light had been spread out into a thick line of different colors, now that line was compressed back into a circle of white light.

Apparently, white light is composed of a vast assemblage of different varieties of light, each with its own characteristic way of being refracted. The group of varieties of light that were least refracted gave rise to the sensation of red; the next group, slightly more refracted, to the sensation of orange; and so on to the most refrangible varieties, which seemed violet.

White light, because of this difference in refrangibility of its components, always breaks into color on passing obliquely from one medium into another of different index of refraction. However, if the second medium is bounded by parallel interfaces (as an ordinary sheet of glass is) the effect produced on entering is canceled on leaving. For that reason white light entering a sheet of glass is white once more on leaving. When the edges of a transparent medium are not parallel, as in a prism, at the edge of a sheet of glass, or in the case of round water droplets, the color production process is not canceled, and a spectrum or a rainbow results.

This means that in determining the index of refraction of a transparent substance, the use of white light introduces an uncertainty, since a whole sheaf of indices of refraction is produced by the various colors present in the light. For that reason it is customary to make use of a particular colored light* in determining indices of refraction. One device frequently used is a "sodium lamp," a device in which light is emitted by the heated sodium vapor within the bulb. This light is yellow in color and is refracted by an amount that varies only over a very small range.

---

* The particular color of light can be defined by its "wavelength." This will be discussed in the next chapter.

By this view of light, it is easy to explain colored objects. It is not necessary to suppose that objects must be either fully transparent (transmitting all colors of light) or fully opaque (transmitting none of them). Some substances may well be opaque to some colors and transparent to others. Red glass, for instance, may possess a chemical admixture that will absorb colors other than red and allow red to pass through. In that case, white light passing through red glass becomes red in color, not because it has gained an impurity from the glass, but only because it has lost to the glass all its components but red. In the same way, an object may reflect some colors and absorb others, appearing colored to the eye for that reason.

It must not be supposed, however, that all yellow objects reflect only yellow light, or that all blue glasses transmit only blue light. It is important to distinguish between physical color and physiological color. A physical color can be identified by the amount of refraction it undergoes in passing from one substance into another. The physiological color is what our brain interprets it to be. The physiological mechanism in the retina of the eye works in such a fashion that a physical orange will give rise to the sensation of orange; therefore it will be a physiological orange as well. However, the retina may be activated in the same way by a mixture of colors that do not include physical orange—for instance, by a mixture of red and yellow. This mixture will then be physiological orange, too.

Light that is colored by transmission through colored glass or reflection from a colored surface need not actually contain the physical colors that correspond to the physiological ones we see. We can determine the physical colors present by passing the light through a prism; for the physiological colors our sight is sufficient, provided, of course, our color vision is normal.

In 1807, the English scientist Thomas Young (1773–1829) pointed out that red, green and blue could, in proper combination, give rise to the sensation of any other color. This was later amplified by the German physiologist Hermann Ludwig Ferdinand von Helmholtz (1821–1894) and is therefore called the Young-Helmholtz theory of color vision.

Many physiologists think that this ability of red, green and blue to create the entire spectrum is a reflection of the situation in the retina of the eye—that is, that there may be three types of color-sensitive retinal cells, one reacting most strongly to red, one to green, and one to blue. The extent to which a particular color of the spectrum or a particular mixture of colors activates each of the three gives rise, therefore, to the color sensation of which we become aware. Light that activates all three equally might then be interpreted as "white"; light that activates the three in one fixed ratio might be "yellow," in another "violet," and so on.

This is made use of in color photography. In one process, the film consists of triple layers, one of which contains a red-absorbing dye so that it is particularly sensitive to red light, another a dye sensitive to blue light, and the third a dye sensitive to green light. The light at each point of the image affects the three in a particular ratio and upon development produces at every point of the image a dye combination in a particular ratio of intensity. The dye combination affects the three pigments of our retina accordingly, and we see the color in the photograph as we would have seen it in the object itself.

Again, a colored print may be obtained by combining dots of a few different colors. Any color can be reproduced by varying the ratio of the colored dots represented. Under a magnifying glass the individual dots may be large enough to be seen in their true color, but if the individual dots cannot be resolved by the unaided eye, neighboring dots will affect the same retinal area and produce a combination of effects that result in the sensation of a color not actually in the dots themselves.

A similar situation is to be found on the screen of a color television set. The screen is covered by an array of dots, some of which react to light by shining blue, some by shining green, and some by shining red. Each particular portion of the TV picture scanned and transmitted by the camera activates these dots in a particular brightness ratio, and we sense that ratio as the same color that was present in the original object.*

Reflecting Telescopes

The fact that white light is a mixture of colors explained what had been observed as an annoying imperfection of the telescope. A parallel beam of light rays passing through a converging lens is brought to a focus on the other side of the lens. The exact position of this focus depends upon the extent to which the light is refracted on passing through the lens, and this at once introduces a complication, since white light consists of a mixture of colors, each

---

* This three-color system is not universally accepted as actually describing the method by which the human eye detects color, but it can certainly be used to produce color photographs and color television.

with its own refractivity, and it is almost always white light we are passing through the lenses of telescopes and microscopes.

The red component of white light is least refracted on passing through the lens and comes to a focus at a particular point. The orange component being refracted to a somewhat greater extent comes to a focus at a point somewhat closer to the lens than the red light. The yellow light is focused closer still, and next follow green, blue and, closest of all, violet. This means that if the eye is so placed at a telescope eyepiece that the red component of the light from a heavenly body is focused upon the retina, the remaining light will be past its focal point and will be broader and fuzzier. The image of the heavenly body will be circled with a bluish ring. If the eye is placed so that the violet end of the spectrum is focused, the remaining light will not yet have reached its focus and there will be an orange rim. The best that can be done is to focus the eye somewhere in the center and endure the colored rims, which are in this way minimized but not abolished.

This is called *chromatic aberration*, "chromatic" coming from a Greek word for color. It would not exist if light were taken from only a small region of the spectrum (such light would be *monochromatic* or "one color"), but a telescope or microscope must take what it gets—usually not monochromatic light.

Newton felt that chromatic aberration was an absolutely unavoidable error of lenses and that no telescope that depended on images formed by the refraction of light through a lens (hence, a *refracting telescope*) would ever be cleared of it. He set about correcting the situation by substituting a mirror for a lens. As was pointed out earlier in the book, a real image is formed by a concave mirror reflecting light, as well as by a convex lens transmitting light. Furthermore, whereas different colors of light are refracted through lenses by different amounts, all are reflected from mirrors in precisely the same way. Therefore, mirrors do not give rise to chromatic aberration.

In 1668, Newton devised a telescope making use of such a mirror. It was the first practical *reflecting telescope*. It was only six inches long and one inch wide, but it was as good as Galileo's first telescope. Shortly thereafter, Newton built larger and better reflecting telescopes.

In addition to the lack of chromatic aberration, reflecting telescopes have additional advantages over refracting telescopes. A lens must be made of flawless glass with two curved surfaces, front and back, ground to as near perfection as possible, if the faint light of the stars is to be transmitted without loss and focused with precision. However, a mirror reflects light, and for this only the reflecting surface need be perfect. In a telescopic mirror it is the forward end that is covered with a thin, reflecting metallic film (not the rear end as in ordinary mirrors), so the glass behind the forward reflecting surface may be flawed and imperfect. It has nothing to do with the light; it is merely the supporting material for the metallized surface in front. Since it is far easier to get a large piece of slightly flawed glass than a large piece of perfect glass, it is easier to make a large telescopic mirror than a large telescopic lens—particularly since only one surface need be perfectly ground in a mirror, rather than two as in the case of the lens.

Again, light must pass through a lens and some is necessarily absorbed. The larger and thicker the lens, the greater the absorption. On the other hand no matter how large a mirror may be, the light is merely reflected from the surface, and virtually none is lost by absorption. Then too a lens can only be supported about the rim, since all other parts must be open to unobstructed passage of light; it becomes difficult to support a large, thick lens about the rim, for the center sags and this introduces distortion. The mirror on the other hand can be supported at as many points as may be desired.

The result is that all the large telescopes in the world are reflectors. The largest currently in operation is the 200-inch reflector, which went into operation in 1948 at Mount Palomar, California. Then there are the 120-inch reflector at Mount Hamilton and the 100-inch reflector at Mount Wilson, both in California. The Soviet Union has recently put a 103-inch reflector into use in the Crimea and has a 236-inch reflector under construction.

Compare this with the 40-inch refractor at Yerkes Observatory in Wisconsin, which has been the largest refractor in use since 1897 and is likely to remain so.

Nevertheless, even the reflectors have by and large reached their practical limit of size. The gathering and concentration of light implies a gathering and concentration of the imperfections of the environment—the haze in the air, the scattered light from distant cities, the temperature variations that introduce rapid variations in the refractivity of air and set the images of the stars to dancing and blurring.

For the next stage in optical telescopy, we may have to await the day (perhaps not very far off) when an astronomical observatory can be set up on the moon, where there is no air to absorb, refract, and scatter the dim light of the stars, and where an astronomer (given the means for survival in a harsh environment) may well consider himself figuratively, as well as literally, in heaven.

But Newton was wrong in thinking that chromatic aberration in lenses was unavoidable. It did not occur to him to test prisms made of different varieties of glass in order to see whether there were the same differences in refraction of the colors of light in all of them. What's more, he ignored the reports of those who did happen to test the different varieties (even Homer nods!).

The difference in degree of refraction for light at the red end and at the violet end of the spectrum determines the degree to which a spectrum is spread out at a given distance from the prism. This is the *dispersion* of the spectrum. The dispersion varies with different types of glass. Thus, flint glass (which contains lead compounds (has a dispersion twice as great as crown glass (which does not contain lead compounds).

One can therefore form a converging lens of crown glass and add to it a less powerful diverging lens of flint glass. The diverging lens of flint glass will only neutralize part of the convergent effect of the crown glass lens, but it will balance all the dispersion. The result will be a combined lens not quite as convergent as the crown glass alone, but one that does not produce a spectrum or suffer from chromatic aberration. It is an *achromatic lens* (from Greek words meaning "no color"). The English optician John Dollond (1706–1761) produced the first achromatic refracting telescope in 1758. While it did not remove all the disabilities of refractors, it did make moderately large refractors practical.

The development of achromatic lenses was of particularly great importance to microscopy. There it was not practical to try to substitute mirrors for lenses and reflection for refraction. For that reason, microscopists had to bear with detail-destroying chromatic aberration long after telescopists had been able to escape.

Through the efforts of the English optician Joseph Jackson Lister (1786–1869) and the Italian astronomer Giovanni Batista Amici (1786?–1863), microscopes with achromatic lenses were finally developed in the early nineteenth century. It was only thereafter that the smaller microorganisms could be seen clearly and that the science of bacteriology could really begin to flourish.

Spectral Lines

Actually, we must not think of sunlight as being composed of a few different colors, as though it were a mixture of seven pigments. Sunlight is a mixture of a vast number of components separated by very slight differences in refractivity. For example, the red portion of the spectrum is not a uniform red but shades imperceptibly into orange.

In the rainbow and in simple spectra such as those that Newton formed, the light seems to be continuous, as though all the infinite possible refractivities were present in sunlight. This is an illusion, however.

If a beam of light passes through a small hole in a blind, let us say, and then through a prism, a large number of circular images are formed, each imprinted in a variety of light of particular refractivity. These overlap and blend into a spectrum. If light of a certain refractivity were missing, neighboring images in either direction would overlap the spot where the missing refractivity ought to have been, and no gap would be visible.

The situation would be improved if the beam of light were made to pass through a narrow slit. The spectrum would then consist of a myriad of images of the slit, each overlapping its neighbor only very slightly. In 1802, the English chemist William Hyde Wollaston (1766–1828) did see a few dark lines in the spectrum, representing missing slit-images. He, however, felt they represented the boundary lines between colors and did not follow through.

Between 1814 and 1824, however, a German optician, Joseph von Fraunhofer (1787–1826), working with particularly fine prisms, noticed hundreds of such dark lines in the spectrum. He labeled the most prominent ones with letters from A to G and carefully mapped the relative position of all he could find. These *spectral lines* are, in his honor, sometimes called *Fraunhofer lines*.

Fraunhofer noticed that the pattern of lines in sunlight and in the light of reflected sunlight (from the moon or from Venus, for instance) was always the same. The light of stars, however, would show a radically different pattern. He studied the dim light of heavenly objects other than the sun by placing a prism at the eyepiece of a telescope, and this was the first use of a *spectroscope*.

Fraunhofer's work was largely disregarded in his lifetime, but a generation later the German physicist Gustav Robert Kirchhoff (1824–1887) put the spectroscope to use as a chemical tool and founded the science of *spectroscopy*.

It was known to chemists that the vapors of different elements heated to incandescence produced lights of different colors. Sodium vapor gave out a strongly yellow light; potassium vapor a dim violet light; mercury a sickly greenish light, and so on. Kirchhoff passed such light through a spectroscope and found that the various elements produced light of only a few refractive varieties. There would only be a few images of the slit, spread widely apart, and this would be an *emission spectrum*. The exact position of each line could be measured against a carefully ruled background, and it could then be shown that each element always produced lines of the same color in the same place, even when it was in chemical combination with other elements. Furthermore, no two elements produced lines in precisely the same place.

The emission line spectrum could be used as a set of "fingerprints" for the elements, therefore. Thus, in 1859, Kirchhoff and his older collaborator, the German chemist Robert Wilhelm Bunsen (1811–1899), while heating a certain mineral to incandescence and studying the emission spectrum of the vapors evolved, discovered lines that did not correspond to those produced by any known element. Kirchhoff and Bunsen therefore postulated the existence of a new element, which they called *cesium* (from the Latin word for "sky blue," because of the sky-blue color of the brightest of the new lines they had observed). The next year they made a similar discovery and announced *rubidium* (from a Latin word for "dark red"). The existence of both metals was quickly confirmed by older chemical techniques.

Kirchhoff observed the reverse of an emission spectrum. Glowing solids emit light of all colors, forming a *continuous spectrum*. If the light of a carbon-arc, for instance, representing such a continuous spectrum is allowed to pass through sodium vapor that is at a temperature cooler than that of the arc, the sodium vapor will absorb some of the light. It will, however, absorb light only of particular varieties—precisely those varieties that the sodium vapor would itself emit if it were glowing. Thus, sodium vapor when glowing and emitting light produces two closely spaced yellow lines that make up virtually the whole of its spectrum. When cool sodium vapor absorbs light out of a continuous spectrum, two dark lines are found to cross the spectrum in just the position of

Portion of solar spectrum in yellow region

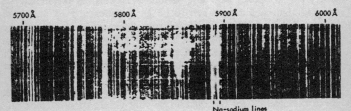

Na-sodium lines

the two bright lines of the sodium emission spectrum. The dark lines represent the sodium *absorption spectrum*.

The dark lines in the solar spectrum seem to be an absorption spectrum. The blazing body of the sun is sufficiently complex in chemical nature to produce what is essentially a continuous spectrum. As the light passes through the somewhat cooler atmosphere, it is partially absorbed. Those parts that would be most strongly absorbed, and that would show up as dark lines in the spectrum, would correspond to the emission spectra of the elements most common in the solar atmosphere. Thus, there are prominent sodium absorption lines in the solar spectrum (Fraunhofer labeled them the "D line"), and this is strong evidence that sodium exists in the solar atmosphere.

In this way, a variety of elements were located in the sun. Indeed, one element, helium, was located in the sun a generation before it was found to exist on the earth. Even the composition of the distant stars could now be determined. While the details of the spectroscopic investigation of the heavens are more appropriately given in a book on astronomy, it might be well to say, in summary, that it was clearly shown that the heavenly bodies are made up of the same chemical elements that make up the earth—though not necessarily in the same proportions.

It also pointed up the dangers of setting limits to human endeavor. The French philosopher Auguste Comte (1798–1857), in an attempt to give an example of an absolute limit set upon man's knowledge, said that it would be forever impossible for man to learn of what material the stars were composed. Had he lived but a few years longer, he would have seen his absolute limit easily breached.

Diffraction

The discovery that white light was actually a mixture of many colors opened new and serious problems for physicists. As long as light could be taken to be an undifferentiated and pure phenomenon, geometrical optics was sufficient. Lines could be drawn representing light rays, and the phenomena of reflection and refraction could be analyzed without any consideration of what the nature of light might be. That question might be left for philosophers.

With light a mixture of colors, it became necessary to seek some explanation for the manner in which light of one color differed from light of another. For that, the question of the nature of light in general had to be considered, and thus was born *physical optics*.

As was pointed out at the start of the book, there are two ways, in general, of avoiding the problem of action at a distance. One is to suppose particles streaming across a space that might then be considered as empty; the other is to suppose waves being propagated through a space that is not really empty. Both types of explanation were advanced for light in the latter half of the seventeenth century.

The more direct of the two alternatives is the particle theory, which Newton himself supported. To begin with, this at once explains the rectilinear propagation of light. Suppose luminous objects are constantly firing tiny particles outward in all directions. If these particles are considered to be massless, a luminous body does not lose weight merely by virtue of being luminous, and light itself would not be affected by gravity. Light, when traveling in an unobstructed path, if unaffected by gravitational force, must travel in a straight path at a constant velocity, as required by Newton's first law of motion (see page I–24). The particles of light would be stopped and absorbed by opaque barriers, and speeding past the edge of the barrier, would cast a sharp boundary between the illuminated area beyond and the barrier-shaded area.

To Newton, the alternative of a wave theory seemed untenable. The wave forms that were familiar to scientists at the time were water waves and sound waves (see page I–156), and these do not necessarily travel in straight lines or cast sharp shadows. Sound waves curve about obstacles, as we know whenever we hear a sound around a corner; and water waves visibly bend about an obstacle such as a floating log of wood. It seemed reasonable to suppose that this behavior was characteristic of wave forms in general.

And yet the particle theory had its difficulties, too. Beams of light could cross at any angle without affecting each other in direction or color, which meant that the light particles did not seem to collide and rebound as ordinary particles would be expected to do. Furthermore, despite ingenious suggestions, there was no satisfactory explanation as to why some light particles gave rise to red sensations, some to green sensations, and so on. The particles had to differ among themselves, of course, but how?

Some of Newton's contemporaries, therefore, took up the wave theory Newton had discarded. The most vigorous supporter of the wave theory in the seventeenth century was the Dutch physicist Christiaan Huygens (1629–1695). He had no real evidence in favor of waves, but he bent his efforts to show that waves could be so treated as to fit the facts of geometric optics. In 1678, he suggested that when a wave front occupies a certain line, each point on the front acts as a source of circular waves, expanding outward indefinitely. These waves melt together, so to speak, and a line can be drawn tangent to the infinite number of little circles centering about each point on the original wave front. This tangent is a picture of the new wave front, which serves as the starting region for another infinite set of circular waves to which another overall tangent can be drawn, and so on.

If waves are analyzed in this fashion, through use of what is now called *Huygen's principle*, it can be shown that a wave front will travel forward in a straight line (at least if only its middle portion is considered) to be reflected with an angle of reflection equal to the angle of incidence, and so on. Furthermore, the waves themselves would have no mass and would be in virtually infinite supply, after the fashion of water waves and sound waves. Being nonmaterial, these light waves would not affect each other upon crossing (and indeed water waves and sound waves can cross each other without interference).

It seemed, then, that there was much to be said for and against each theory. One must therefore look for places where the two theories differ as to the nature of the phenomena they predict. Through an observation of conditions under which such phenomena should exist, one or the other theory (or conceivably both) may be eliminated. (This is the method generally used wherever theories conflict or overlap.)

For instance, Huygens' wave theory could explain refraction under certain conditions. Suppose a straight wave front of light strikes the plane surface of glass obliquely. One end of the wave front strikes the glass first, but suppose its progress is slowed as it enters the glass. In that case, when the next section of the front hits the glass, it has gained on the first section, for the second has been traveling through air, while the first has been traveling, more slowly, through glass. As each section of the wave front strikes, it is slowed and gained upon by the portion of the wave front that has not yet struck. The entire wave front is in this way refracted and, in entering the glass, makes a smaller angle with the normal. On emerging from the glass, the first section to emerge speeds up again and gains on those portions that have not yet emerged. The emerging light takes on its original direction again.

An analogy can be drawn between this and a line of marching soldiers leaving a paved highway obliquely and entering a plowed field. The soldiers leaving the highway are, naturally, slowed down; those first to enter the field are slowed down first, and the whole line of soldiers (if they make no effort to co ct for the change in footing) must alter the direction of march toward the direction of the normal to the highway-field interface.

Thus, the wave theory can explain refraction by supposing that the velocity of light is less in glass than in air. By making a further assumption, it can also explain spectrum formation. If light is a wave form, it must have a *wavelength* (the length from the crest of one wave to that of another, see page I–152). Suppose, then, that this wavelength varies with color, being longest at the red end of the spectrum and shortest at the violet end. It would seem reasonable then to suppose that short wavelengths are slowed more sharply on entering glass from air than are long wavelengths. (Again as an analogy, a marching soldier with a short stride would sink into the plowed field more times in covering a certain distance than would another soldier with a long stride. The short-striding soldier would then be slowed down more, and the marching line of soldiers—if no effort were made to correct matters—would break up into groups marching in slightly different directions depending on the length of their stride.)

In short, red light would be least refracted, orange next, and so on. In this way, light passing through a prism would be expected to form a spectrum.

Newton could explain refraction by his particle theory, too, but he was forced to assume that the velocity of the particles of light increased in passing from a medium of low optical density to one of high optical density. Here, then, was a clear-cut difference in the two theories. One had only to measure the velocity of light in different media and note the manner in which that velocity changed; one could then decide between the Newtonian particles and the Huygensian waves. The only catch was that it was not until nearly two centuries after the time of Newton and Huygens that such a measurement could be made (see page 74).

However, there was a second difference in the predictions of the theories. Newton's light particles traveled in straight lines in all portions of a light beam, so the beam might be expected to cast absolutely sharp shadows. Not so, Huygens' waves. Each point in the wave front served as a focus for waves in all directions, but through most of the wave front, a wave to the right from one point was canceled by a wave to the left from the neighboring point on the right, and so on. After all cancellations were taken into account, only the forward motion was left. There was an exception, however, at the ends of the wave front. At the right end, a rightward wave was not canceled because there was no rightward neighbor to send out a leftward wave. At the left end, a leftward wave was not canceled. A beam of light, therefore, had to "leak" sideways if it was a wave form. In particular, if a beam of light passed through a gap in an opaque barrier, the light at the boundary of the beam, just skimming the edge of the gap, ought to leak sideways so that the illuminated portion of a surface farther on ought to be wider than one would expect from strictly straight-line travel.

This phenomenon of a wave form bending sideways at either end of a wave front is called *diffraction,* and this is, in fact, easily observed in water waves and sound waves. Since light, on passing through a gap in a barrier, did not seem to exhibit diffraction, the particle theory seemed to win the nod. Unfortunately, what was not clearly understood in Newton's time was that the smaller the wavelength of any wave form, the smaller the diffraction effect. Therefore, if one but made still another assumption—that the wavelength of light waves was very small—the diffraction effect would be expected to be very hard to observe, and a decision might still be suspended.

As a matter of fact, the diffraction of light *was* observed in the seventeenth century. In 1665, an Italian physicist, Francesco Maria Grimaldi (1618?–1663), passed light through two apertures and showed that the final band of light on the surface that was illuminated was a trifle wider than it ought to have been if light had traveled through the two apertures in an absolutely straight fashion. In other words, diffraction had taken place.

What was even more important was that the boundaries of the illuminated region showed color effects, with the outermost portions of the boundary red and the innermost violet. This, too, it was eventually understood, fit the wave theory, for if red light had the longest wavelengths it would be most diffracted, while violet light, with the shortest wavelengths, would be least diffracted.

Indeed, this principle came to be used to form spectra. If fine parallel lines are scored on glass, each will represent an opaque region separated by a transparent region. There will be a series of gaps, at the edges of which diffraction can take place. In fact, if the gaps are very narrow, the glass will consist entirely of gap edges, so to speak. If the scoring is very straight and the gaps are very narrow, the diffraction at each edge will take place in the same fashion, and the diffraction at any one edge will reinforce the diffraction at all the others. In this way, a spectrum as good as, or better than, any that can be formed by a prism will be produced. Lines can be scored more finely on polished metal than on glass. In such a case, each line is an opaque region separated by a reflecting region, and this will also form a spectrum (though ordinary reflection from unbroken surfaces will not).

The spectra formed by such *diffraction gratings* are reversed in comparison to spectra formed by refraction. Where violet is most refracted and red least, violet is least diffracted and red most. Consequently, if the spectrum in one case is "red-left-violet-right," it is "red-right-violet-left" in the other. More exactly, in the case of refraction spectra, red is nearest the original line in which light was traveling, while violet is nearest the original line in the case of diffraction spectra.

(Nowadays, diffraction gratings are used much more commonly than prisms in forming spectra. The first to make important use of diffraction gratings for this purpose was Fraunhofer, the man who first made thorough observations of spectral lines.)

Newton was aware of Grimaldi's experiments and even repeated them, particularly noting the colored edges. However, the phenomenon seemed so minor to him that he did not feel he could suspend the particle theory because of it, and so he disregarded its significance. More dramatic evidences of diffraction, and the ability to measure the velocity of light in different media, still remained far in the future. What it amounted to, then, was that physicists of the seventeenth century had to choose between two personalities rather than between two sets of physical evidence. Newton's great prestige carried the day, and for a hundred years afterward, throughout the eighteenth century, light was considered by almost all physicists to be indisputably particulate in nature.

CHAPTER 5

Light Waves

Interference

The eighteenth century confidence in the existence of light particles came to grief at the very opening of the nineteenth century. In 1801, Young (of the Young-Helmholtz theory of color vision) conducted an experiment that revived the wave theory most forcefully.

Young let light from a slit fall upon a surface containing two closely adjacent slits. Each slit served as the source of a cone of light, and the two cones overlapped before falling on a screen.

If light is composed of particles, the region of overlapping should receive particles from both slits. With the particle concentration therefore doubled, the overlapping region should be uniformly brighter than the regions on the outskirts beyond the overlapping, where light from only one cone would be received. This proved not to be so. Instead, the overlapping region consisted of stripes—bright bands and dim bands alternating.

For the particle theory of light, this was a stopper. On the basis of the wave theory, however, there was no problem. At some points on the screen, light from both the first and second cone would consist of wave forms that were *in phase* (that is, crest matching crest, trough matching trough—see page I–151). The two light beams would reinforce each other at those points so that there would be a resultant wave form of twice the amplitude and, therefore, a region of doubled brightness. At other points on the screen, the two light beams would be *out of phase* (with crest matching trough and trough matching crest). The two beams would then cancel, at least in part, and the resultant wave form would have a much smaller amplitude than either component; where canceling was perfect, there would be no wave at all. A region of dimness would result.

In short, whereas one particle of the type Newton envisaged for light could not interfere with and cancel another particle, a wave form can and does easily interfere with and cancel another wave form. *Interference patterns* can easily be demonstrated in water waves, and interference is responsible for the phenomenon of beats; for instance (see page I–150), in the case of sound waves. Young was able to show that the wave theory would account for just such an interference as was observed.

Furthermore, from the spacing of the interference bands of light and dark, Young could calculate the wavelength of light. If the ray of light from one cone is to reinforce the ray of light from the second cone, both rays must be in phase, and that means the distances from the point of reinforcement on the screen to one slit and to the other must differ by an integral number of wavelengths. By choosing the interference bands requiring the smallest difference in distances, Young could calculate the length of a single wavelength and found it to be of the order of a fifty-thousandth of an inch, certainly small enough to account for the difficulty of observing diffraction effects (see page 64). It was possible to show, furthermore, that the wavelengths of red light are about twice the wavelengths of violet light, which fit the requirements of wave theory if spectrum formation is to be explained.

In the metric system, it has proved convenient to measure the wavelengths of light in *millimicrons* ($m\mu$),* where a milli-micron is a billionth of a meter (10^{-9}m) or a ten-millionth of a centimeter (10^{-7}cm). Using this unit, the spectrum extends from 760 $m\mu$ for the red light of longest wavelength to 380 $m\mu$ for the violet light of shortest wavelength. The position of any spectral line can be located in terms of its wavelength.

One of those who made particularly good measurements of the wavelengths of spectral lines was the Swedish astronomer and physicist Anders Jonas Ångström (1814–1874), who did his work

* The Greek letter "mu" (μ) commonly symbolizes a "micrometer" or "micron" which is one-millionth of a meter.

in the mid-nineteenth century. He made use of a unit that was one-tenth of a millimicron. This is called an *angstrom unit* (A) in his honor. Thus, the range of the spectrum is from 7600 A to 3800 A.

The wavelength ranges for the different colors may be given roughly (for the colors blend into one another and there are no sharp divisions) as: red, 7600–6300 A; orange, 6300–5900 A; yellow, 5900–5600 A; green, 5600–4900 A; blue, 4900–4500 A; violet 4500–3800 A.

Incandescent sodium vapor gives off a bright line in the yellow, while sodium absorption produces a dark line in the same spot. This line, originally considered to be single and given the letter D by Fraunhofer has, by the use of better spectroscopes, been resolved into two very closely spaced lines, D_1 and D_2. The former is at wavelength 5896 A, the latter at 5890 A. Similarly, Fraunhofer's C line (in the red) and F line (in the blue) are both produced by hydrogen absorption and have wavelengths of 6563 A and 4861 A respectively. (Indeed, it was Angström who first showed from his study of spectral lines that hydrogen occurred in the sun.) In similar fashion, all the spectral lines produced by any element, through absorption or emission, can be located accurately.

The wave theory of light was not accepted at once despite the conclusiveness (in hindsight) of Young's experiment. However, all through the nineteenth century, additional evidence in favor of light waves turned up, and additional phenomena that, by particle theory, would have remained puzzling, found ready and elegant explanations through wave theory. Consider, for instance, the color of the sky. . . .

Light, in meeting an obstacle in its otherwise unobstructed path, undergoes a fate that depends on the size of the obstacle. If the obstacle is greater than 1000 mμ in diameter, light is absorbed and there is an end to a light ray, at least in the form of light. If the obstacle is smaller than 1 mμ in diameter, the light ray is likely to pass on undisturbed. If, however, the obstacle lies between 1 mμ and 1000 mμ in diameter, it will be set to vibrating as it absorbs the light and may then emit a light ray equal in frequency (and therefore in wavelength) to the original, but traveling in a different direction. This is *light scattering*.

The tiny water or ice particles in clouds are of a size to scatter light in this fashion; therefore a cloud-covered sky is uniformly white (or, if the clouds are thick enough to absorb a considerable fraction of the light altogether, uniformly gray).

The dust normally present in the atmosphere also scatters light. Shadows are therefore not absolutely black but, although darker by far than areas in direct sunlight, receive enough scattered light to make it possible to read newspapers in the shade of a building or even indoors on a cloudy day.

After the sun has set, it is shining past the bulge of the earth upon the upper atmosphere. The light scattered downward keeps the earth in a slowly dimming *twilight*. It is only after the sun has sunk 18° below the horizon that full night can be said to have begun. In the morning, sunrise is preceded by a second twilight period which is *dawn*.

As particles grow smaller, a pronounced difference becomes noticeable in the amount of scattering with wavelength. The light of short wavelength is scattered to a greater extent than the light of long wavelength. Thus, if sunlight shines down upon a cloud of tobacco smoke, it is the short-wave light that is more efficiently scattered, and the tobacco smoke therefore seems bluish.

The British physicist John Tyndall (1820–1893) studied this phenomenon. He found that light passing through pure water or a solution of small-molecule substances such as salt or sugar, underwent no scattering. The light beam, traveling only forward, cannot be seen from the side and the liquid is *optically clear*. If the solution contains particles large enough to scatter light, however (examples are the molecules of proteins or small conglomerates of ordinarily insoluble materials such as gold or iron oxide) some of the light is emitted sideways, and the beam can then be seen from the side. This is the *Tyndall effect*.

The English physicist John William Strutt, Lord Rayleigh (1842–1919), went into the matter in greater detail in 1871. He worked out an equation that showed how the amount of light scattered by molecules of gas varied with a number of factors, including the wavelength of the light. He showed the amount of scattering was inversely proportionate to the fourth power of the wavelength. Since the red end of the spectrum had twice the wavelength of the violet end, the red end was scattered less (and the violet end scattered more) by a factor of 2^4, or 16.

Over short distances, the scattering by particles as small as gas molecules of the atmosphere is insignificant. If, however, the miles of atmosphere stretching overhead are considered, scattering mounts up and, as Rayleigh showed, must be confined almost entirely to the violet end of the spectrum. Enough light is so scattered to drown out the feeble light of the stars (which are, of course, present in the sky by day as well as by night). Furthermore, the scattered light that illuminates the sky, heavily represented in the short-wave region, is blue in color; the sun itself, with that small quantity of shortwave light substracted from its color, is a trifle redder than it would be if the atmosphere were absent.

This effect is accentuated when the sun is on the horizon, for it then shines through a greater thickness of air as its light comes obliquely through the atmosphere. Enough light is scattered from even the midportions of the spectrum to lend the sky a faintly greenish hue, while the sun itself, with a considerable proportion of its light scattered, takes on a ruddy color indeed. This, reflected from broken clouds, can produce a most beautiful effect. Since the evening sky, after the day's activities, is dustier than is the morning sky, and since the dust contributes to scattering, sunsets tend to be more spectacular than sunrises. After gigantic volcano eruptions (notably that of Krakatoa, which blew up—literally—in 1883) uncounted tons of fine dust are hurled into the upper atmosphere, and sunsets remain particularly beautiful for months afterward.

On the moon (which lacks an atmosphere) the sky is black even when the sun is present in the sky. Shadows are pitch black on the moon, and the terminator (the line between the sunlit and shadowed portion of the body) is sharp, because there is neither dawn nor twilight. The earth, as seen from space, would also possess a terminator, but a fuzzy one that gradually shaded from light to dark. Furthermore, its globe would have a distinctly bluish appearance, thanks to light scattering by its atmosphere.

The Velocity of Light

In time, even the question of the velocity of light in various media was settled in favor of Huygens' view as one climax to two centuries of work on the problem. The first effort to measure the velocity of light was made by Galileo about a half-century before the wave-particle controversy began.

Galileo placed himself on a hilltop and an assistant on another about a mile away. It was his intention to flash a lantern in the night and have his assistant flash a lantern in return as soon as he spied Galileo's light. The time lapse between Galileo's exposure of light and his sighting of the return signal would then, supposedly, represent the time it took for light to travel from Galileo to the assistant and back. About that same period, this principle was also used successfully in determining the velocity of sound (see page I–164).

Galileo found a perceptible delay between the light emission and return; however, it was obvious to him that this was due not to the time taken by light to travel but to that taken by the human nervous system to react to a sensation, for the delay was no longer when the two men were a mile apart than when they were six feet apart.

Consequently, all Galileo could show by his experiment was that light traveled far more rapidly than sound. In fact, it remained possible that light traveled with infinite speed, as indeed many scholars had surmised.

It was not until the 1670's that definite evidence was presented to the effect that the velocity of light, while very great, was, nevertheless, finite. The Danish astronomer Olaus Roemer (1644–1710) was then making meticulous observations of Jupiter's satellites (which had been discovered by Galileo in 1610). The orbits of those satellites had been carefully worked out, and the moments at which each satellite ought to pass behind Jupiter and be eclipsed to the sight of an observer on earth could, in theory, be calculated with precision. Roemer found, however, that the eclipses took place off schedule, several minutes too late at some times and several minutes too early at others.

On further investigation, he discovered that at the times that earth and Jupiter were on the same side of the sun, the eclipses

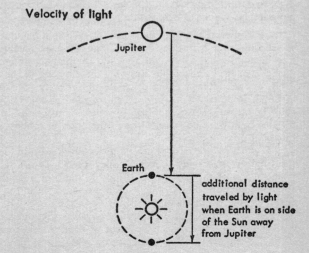

Velocity of light

Jupiter

Earth

additional distance traveled by light when Earth is on side of the Sun away from Jupiter

were ahead of schedule; when the two planets were on the opposite sides of the sun, they were behind schedule.

Imagine a beam of light connecting a Jovian satellite with the earth—that is, a beam by means of which we see the satellite. At the moment of eclipse, the beam is cut off, and we no longer see the satellite. At least that would be the situation if light traveled with an infinite velocity. As soon as the beam was cut off, it would, under those conditions, cease to exist all along its path of travel, however long that path might be. It would not matter whether earth was on the same side of the sun as Jupiter was, or on the opposite side.

If, however, light traveled at a finite velocity, then once the beam was cut off by Jupiter, light would continue to travel onward toward earth; the earth observer would therefore continue to see the satellite until such time as the "broken end" of the beam of light reached him. Then, and only then, would the satellite disappear in eclipse. There would be a finite time between the actual eclipse and the eclipse as seen. The greater the distance between Jupiter and the earth, the greater this time lapse.

If the distance separating Jupiter and the earth were always the same, this time lapse would be constant and might, therefore, be ignored. But the distance between Jupiter and the earth is not constant. When earth and Jupiter are on the same side of the sun, they are separated by as little as 400,000,000 miles. When they are on opposite sides, they can be separated by that distance plus the full width of the earth's orbit, or a total of about 580,000,000 miles. If at closest approach the eclipse is, say, eight minutes ahead of schedule, and at furthest distance, eight minutes behind schedule, then it would take light roughly 16 minutes to cross the diameter of the earth's orbit. Knowing the diameter of the earth's orbit, one could then easily calculate the velocity of light and, in 1676, Roemer did so. In the light of modern knowledge, the value he obtained was rather low. However, he succeeded in showing that light traveled at velocities of the order of a hundred fifty thousand miles a second.

Roemer's work was not accepted wholeheartedly, but in 1728 the English astronomer James Bradley (1693–1762) used the phenomenon of the *aberration of light* to perform a similar calculation. Suppose that the light of a star from near the North Celestial Pole is descending vertically upon the earth. The earth, however, is moving in its orbit at right angles to that direction and is therefore moving to meet the beam of light. A telescope must be angled slightly to catch that beam of light, just as an umbrella must be angled slightly to catch the raindrops if you are walking in a storm in which the rain is falling vertically.

The telescope must be angled in a continually changing direction as the earth moves about its curved orbit; therefore, the star seems to move in a very tiny ellipse in the sky. The size of the ellipse depends on the ratio of the velocity of the earth's motion to that of light's motion. (There would be no aberration if the earth were standing still or if light traveled at infinite velocity.) Since the earth's velocity around the sun is known to be 18.5 miles per second, the velocity of light could easily be calculated. Bradley showed that the velocity of light was nearly 190,000 miles a second.

It was not until 1849, however, that the question of the velocity of light was brought down from the heavens and measured upon the earth. The experimenter who did this was the French physicist Armand Hippolyte Louis Fizeau (1819–1896), who returned to Galileo's principle but tried to eliminate the matter of human reaction time.

He did this by allowing the light from one hilltop to be returned from another hilltop, not by a human being, but by a mirror. Furthermore, the light being emitted had to pass between the cogs of a turning wheel; consequently, the light was "chopped up" into a series of fragments—a dotted line of light, so to speak.

Consider the behavior of such an interrupted beam. Light travels so rapidly that if the wheel were turning at an ordinary rate of speed, each bit of light emerging between the cogs of the wheel would streak to the mirror, be reflected, and streak back before the wheel had time to move very far. The light would return through the same gap in the cogs by which it had left. A person viewing the returning light through an eyepiece would see a series of light pulses at such short intervals that he would seem to see one continuous blur of light. Furthermore, the light would be quite bright, because almost all the light that was emitted would be returned.

Of course, the last bit of one of the fragments of light, the bit that had just slipped between the cogs as one cog was about to cut off the beam, would find the cog completely in the way when it returned, and it would be absorbed. Consequently, the reflected light would lose just a tiny bit of its intensity and would not be quite as bright as it would be if there were no cogged wheel in the way at all.

If the cogged wheel were made to rotate faster and faster, a larger and larger fraction of the light would be intercepted by the cog on its return, and the reflected light, as seen through the eye-piece, would grow dimmer and dimmer. Eventually, this dimness would reach a minimum, because all the light that emerged while the gap was passing would return while the cog was passing. But if the wheel were rotated still faster, then some of the light would begin slipping through the next gap, and the light would begin to brighten again. At a certain point, all the light passing through one gap would return through the next, and there would be light at maximum brightness again.

By measuring the speed of rotation of the cogged wheel at the time of both minimum and maximum brightness, and knowing the distance from the light source to the mirror, one could cal-culate the speed of light. Fizeau's results were not as accurate as Bradley's, for instance, but Fizeau had brought the measurement to earth and had not involved any heavenly bodies.

Fizeau had a co-worker, the French physicist Jean Bernard Léon Foucault (1819–1868), who introduced an improvement that further eliminated human error. In Fizeau's device, it was still necessary to choose the points at which the brightness of the light seemed at a minimum or at a maximum. This required human judgment, which was unreliable. Foucault introduced a second mirror in place of the toothed wheel. The second mirror was set to turning. The turning mirror sent light to the fixed mirror only when it was turned in the proper direction. By the time the light had been reflected from the fixed mirror, the turning mirror had moved slightly. The returning light was reflected, therefore, not back to the fixed mirror again, but at a slight angle. With little trouble, this slight angle could be measured off a scale. From that, from the rate at which the mirror turned and from the dis-tance between the two mirrors, the velocity of light could be measured with considerable accuracy, and was.

What's more, Foucault was able to make the same measure-ment when light was made to travel through water rather than through air. This he did in 1850 and found that the velocity of light in water was distinctly less than that in air. This was precisely in accord with Huygens' prediction of nearly two centuries before and counter to Newton's prediction. To physicists, this seemed to be the last straw, and there was no important resistance to the wave theory of light after that.

The velocity of light passing through any transparent medium is equal to its velocity in a vacuum divided by the index of refrac-tion (n) of the medium. The velocity of light in a vacuum is customarily represented as *c*, which stands for *celeritas*, a Latin word for "velocity." We might say then:

$$v = \frac{c}{n}$$
(Equation 5-1)

If we accept the approximate value of 186,000 miles per second for *c*, then since the index of refraction of water is 1.33, the velocity of light in water is 186,000/1.33, or 140,000 miles per second. Similarly, the velocity of light in glass with an index of refraction of 1.5 is 124,000 miles per second, while through diamond, with its index of refraction of 2.42, the velocity of light is 77,000 miles per second.

No substance with an index of refraction of less than 1 has been discovered, nor, on the basis of present knowledge, can any such substance exist. This is another way of saying that light travels more rapidly in a vacuum than in any material medium.

Since Foucault's time, many added refinements have been brought to the technique of measuring the velocity of light. In 1923, the American physicist Albert Abraham Michelson (1852–1931) made use of a refined version of Foucault's setup and separated his mirrors by a distance of 22 miles, estimating that distance to an accuracy of within an inch. Still later, in 1931, he decided to remove the trifling interference of air (which has an index of refraction slightly greater than 1 and which carries haze and dust besides) by evacuating a tube a mile long and arranging combinations of mirrors in such a way as to allow the light beam to move back and forth till it had traveled ten miles in a vacuum, all told.

Michelson's last measurement had pinned the velocity down to within ten miles per second of what must be the correct value, but that did not satisfy physicists. In 1905 (as we shall have occasion to see later in the volume, see page 102ff), the velocity of light in a vacuum was revealed to be one of the fundamental constants of the universe, so there could be no resting while it remained possible to determine that velocity with a little more accuracy than had been possible hitherto. Consequently, new and more refined methods for measuring the velocity of light have been brought into use since World War II, and in 1963, the National Bureau of Standards adopted the following value for *c*: 186,281.7 miles per second.

To be precisely accurate, they adopted the value in metric units, and here, by a curious coincidence, the velocity of light comes out to an almost even value: 299,792.8 kilometers per second.

As you see, this is just a trifle short of 300,000 kilometers per second, or 30,000,000,000 centimeters per second. This latter value can be given as 3×10^{10} cm/sec.

At this velocity, light can travel from the moon to the earth in 1 1/4 seconds, and from the sun to the earth in eight minutes. In one year, light travels 9,450,000,000,000 kilometers, or 5,900,000,000,000 miles, and this distance is called a *light-year*.

The light-year has become a convenient unit for use in astronomy since all objects outside our solar system are separated from us by distances so vast that no smaller unit will do. Our nearest neighbors among the stars, the members of the Alpha Centauri system, are 4.3 light years away, while the diameter of our Galaxy as a whole is some 100,000 light-years.

The Doppler-Fizeau Effect

With light viewed as a wave motion, it was reasonable to predict that it would exhibit properties analogous to those shown by other wave motions. The Austrian physicist Johann Christian Doppler (1803–1853) had pointed out that the pitch of sound waves varied with the motion of the source relative to the listener. If a sound-source were approaching the listener, the sound waves would be crowded together, and more waves would impinge upon the ear per second. This would be equivalent to a raised frequency, so the sound would be heard as being of higher pitch than it would have been heard if the source were fixed relative to the listener. By the same reasoning, a receding sound source emits a sound of lower pitch, and the train whistle, as the train passes, suddenly shifts from treble to base (see page I–174).

In 1842, Doppler pointed out that this *Doppler effect* ought to apply to light waves, too. In the case of an approaching light source, the waves ought to be crowded together and become of higher frequency, so the light would become bluer. In the case of a receding light source, light waves would be pulled apart and become lower in frequency, so the light would become redder.*

Doppler felt that all stars radiated white light, with the light more or less evenly distributed across the spectrum. Reddish

---

* Please note that this change does not affect the velocity of light. The wavelength may decrease or increase according to the relative motion of the light source and the observer; however, whether the waves are long, short or in-between, they all move at the same velocity.

stars, he felt, might be red because they were receding from us, while bluish stars were approaching us. This suggestion, however, was easily shown to be mistaken, for the fallacy lies in the assumption that the light we see is all the light there is. . . .

So intimately is light bound up with vision that one naturally assumes that if one sees nothing, no light is present. However, light might be present in the form of wavelengths to which the retina of the eye is insensitive. Thus, in 1800, the British astronomer William Herschel (1738–1822) was checking the manner in which different portions of the spectrum affected the thermometer. To his surprise, he found that the temperature rise was highest at a point somewhat below the red end of the spectrum—a point where the eye could see nothing.

When the wave theory was established, the explanation proved simple. There were light waves with wavelengths longer than 7600 A. Such wavelengths do not affect the eye, and so they are not seen; nevertheless, they are real. Light of such long wavelengths can be absorbed and converted to heat; they can therefore be detected in that fashion. They could be put through the ordinary paces of reflection, refraction, and so on, provided that detection was carried through by appropriate heat-absorbing instruments and not by eye. These light waves, as received from the sun, could even be spread out into a spectrum with wavelengths varying from 7600 A (the border of the visible region) up to some 30,000 A.

This portion of the light was referred to as "heat rays" on occasion, because they were detected as heat. A better name, however, and one universally used now, is *infrared radiation* ("below the red").

The other end of the visible spectrum is also not a true end. Light affects certain chemicals and, for instance, will bring about the breakdown of silver chloride, a white compound, and produce black specks of metallic silver. Silver chloride therefore quickly grays when exposed to sunlight (and it is this phenomenon that serves as the basis for photography). For reasons not understood in 1800, but which were eventually explained in 1900 (see page 132), the light toward the violet end of the spectrum is more efficient in bringing about silver chloride darkening than is light at the red end. In 1801, the German physicist Johann Wilhelm Ritter (1776–1810) found that silver chloride was darkened at a point beyond the violet end of the spectrum in a place where no light was visible. What's more, it was darkened more efficiently there than at any place in the visible spectrum.

Thus, there was seen to be a "chemical ray" region of the spectrum, one more properly called *ultraviolet radiation* ("beyond the violet"), where the wavelength was smaller than 3600 A. Even the early studies carried the region down to 2000 A, and in the twentieth century, far shorter wavelengths were encountered.

By mid-nineteenth century, then, it was perfectly well realized that the spectrum of the sun, and presumably of the other stars, extended from far in the ultraviolet to far in the infrared. A relatively small portion in the middle of the spectrum (in which, however, solar radiation happens to be at peak intensity), distinguished only by the fact that the wavelengths of this region stimulated the retina of the eye, was what all through history had been called "light." Now it had to be referred to as *visible light*. What, before 1800, would have been a tautology, had now become a useful phrase, for there was much invisible light on either side of the visible spectrum.

It can now be seen why Doppler's suggestion was erroneous. The amount of the Doppler shift in any wave form depends on the velocity of the wave form as compared with the velocity of relative motion between wave source and observer. Stars within our Galaxy move (relative to us) at velocities that are only in the tens of kilometers per second, while the velocity of light is 300,000 kilometers per second. Consequently, the Doppler effect on light would be small indeed. There would be only a tiny shift toward either the red or the blue—far from enough to account for the visible redness or blueness of the light of certain stars. (This difference in color arises from other causes, see page 128.)

Furthermore, if there is a tiny shift toward the violet, some of the violet at the extreme end does, to be sure, disappear into the ultraviolet, but this is balanced by a shift of some of the infrared into the red. The net result is that the color of the star does not change overall. The reverse happens if there is a shift to the red, with infrared gaining and ultraviolet losing, but with the overall visible color unchanged.

Fizeau pointed this out in 1848 but added that if one fixed one's attention on a particular wavelength, marked out by the presence of a spectral line, one might then detect its shift either toward the red or toward the violet. This turned out to be so, and in consequence the Doppler effect with respect to light is sometimes called the *Doppler-Fizeau effect*.

Important astronomical discoveries were made by noting changes in the position of prominent spectral lines in the spectra of heavenly bodies, as compared with the position of those same spectral lines produced in the laboratory, where no relative motion is involved. It could be shown by spectral studies alone, for instance, that the sun rotated, for one side of the rotating sun is receding and the other advancing; the position of the spectral lines in the light from one side or the other therefore reflected this. Again, light from Saturn's rings showed that the outer rim was moving so much more slowly than the inner rim that the rings could not be rotating as a single piece and must consist of separate fragments.

In 1868, the English astronomer William Huggins (1824–1910) studied the lines in the spectrum of the star Sirius and was able to show that Sirius was receding from us at a speed of some 40 kilometers per second (a value reduced by later investigations). Since then, thousands of stars have had their *radial velocities* (velocities toward or away from us) measured, and most such velocities fall in the range of 10 to 40 kilometers per second. These velocities are toward us in some cases and away from us in others.

In the twentieth century, such measurements were made on the light from galaxies outside our own. Here it quickly turned out that there was a virtually universal recession. With the exception of one or two galaxies nearest us, there was an invariable shift in spectral lines toward the red—an effect which became famous as the *red shift*. Furthermore, the dimmer (and, therefore, presumably the farther) the galaxy, the greater the red shift. This correlation of distance with velocity of recession would be expected if the galaxies were, one and all, moving farther and farther from each other, as though the whole universe were expanding; this, indeed, is the hypothesis usually accepted to explain the red shift.

Doppler—Fizeau shift

violet   red

light source approaching

light source stationary

light source receding

As the red shift increases with the distance of the galaxies, the velocity of recession, relative to ourselves, increases, too. For the very distant galaxies, these velocities become considerable fractions of the velocity of light. Velocities up to four-fifths the velocity of light have been measured among these receding galaxies. Under such circumstances, there is a massive shift of light into the infrared, a greater shift than can be replaced from the ultraviolet radiation present in the light of these galaxies. The total visible light of such distant galaxies dims for that reason and sets a limit to how much of the universe we might see by visible light, no matter how great our telescopes.

Polarized Light

To say that light consists of waves is not enough, for there are two important classes of waves with important differences in properties. Thus, water waves are *transverse waves*, undulating up and down at right angles to the direction in which the wave as a whole is traveling. Sound waves are *longitudinal waves*, undulating back and forth in the same direction in which the wave as a whole is traveling (see page I-156). Which variety represents light waves?

Until the second decade of the nineteenth century, the scientific minority who felt light to be a wave form believed it to be a longitudinal wave form. Huygens thought this, for instance. However, there remained a seventeenth century experiment on light that had never been satisfactorily explained by either Newton's particles of light or Huygens' longitudinal waves of light, and this eventually forced a change of mind.

The experiment was first reported in 1669 by a Dutch physician, Erasmus Bartholinus (1625-1698). He discovered that a crystal of Iceland spar (a transparent form of calcium carbonate) produced a double image. If a crystal was placed on a surface bearing a black dot, for instance, two dots were seen through the crystal. If the crystal was rotated in contact with the surface, one of the dots remained motionless while the other rotated about it. Apparently, light passing through the crystal split up into two rays that were refracted by different amounts. This phenomenon was therefore called *double refraction*. The ray that produced the motionless dot, Bartholinus dubbed the *ordinary ray;* the other, the *extraordinary ray*.

Both Huygens and Newton considered this experiment but

could come to no clear conclusion. Apparently, if light was to be refracted in two different ways, its constituents, whether particles or longitudinal waves, must differ among themselves. But how?

Newton made some vague speculations to the effect that light particles might differ among themselves as the poles of a magnet did (see page 141). He did not follow this up, but the thought was not forgotten.

In 1808, a French army engineer, Étienne Louis Malus (1775–1812), was experimenting with some doubly refracting crystals. He pointed one of them at sunlight reflected from a window some distance outside his room and found that instead of seeing the shining spot of reflected sunlight double (as he expected) he saw it single. He decided that in reflecting light the window reflected only one of the "poles" of light of which Newton had spoken. The reflected light he therefore called *polarized light*. It was a poor name that did not represent the actual facts, but it has been kept and will undoubtedly continue to be kept.

When the wave theory of light sprang back into prominence with Young's experiment, it soon enough became clear that if light were only considered transverse waves, rather than longitudinal waves, polarized light could easily be explained. By 1817, Young had come to that conclusion, and it was further taken up by a French physicist, Augustin Jean Fresnel (1788–1827). In 1814, Fresnel had independently discovered interference patterns, and he went on to deal with transverse waves in a detailed mathematical analysis.

To see how transverse waves will explain polarization, imagine a ray of light moving away from you with the light waves undulating in planes at right angles to that line of motion, as is required of transverse waves. Say the light waves are moving up and down. They might also, however, move right and left and still be at right angles to the line of motion. They might even be moving diagonally at any angle and still be at right angles to the line of motion. When the component waves of light are undulating in all possible directions at right angles to the line of motion, and are evenly distributed through those planes, we have *unpolarized light*.

Let's concentrate on two forms of undulation, up-down and left-right. All undulations taking up diagonal positions can be divided into an up-down component and a left-right component (just as forces can be divided into components at right angles to each other, see page I–40). Therefore, for simplicity's sake we can consider unpolarized light as consisting of an up-down component and a left-right component only, the two present in equal intensities.

It is possible that the up-down component may be able to slip through a transparent medium where the left-right component might not. Thus, to use an analogy, suppose you held a rope that passed through a gap in a picket fence. If you made up-down waves in the rope, they would pass through the gap unhindered. If you made left-right waves, those waves would collide with the pickets on either side of the gap and be damped out.

The manner in which light passes through a transparent substance, then, depends on the manner in which the atoms making up the substance are arranged—how the gaps between the atomic pickets are oriented, in other words. In most cases, the arrangement is such that light waves in any orientation can pass through with equal ease. Light enters unpolarized and emerges unpolarized. In the case of Iceland spar, this is not so; only up-down light waves and left-right light waves can pass through, and one of these passes with greater difficulty, is slowed up further, and therefore is refracted more. The result is that at the other end of the crystals two rays emerge—one made up of up-down undulations only and one made up of left-right undulations only. Each of these is a ray of polarized light. Because the undulations of the light waves in each of these rays exist in one plane only, such light may more specifically be called *plane-polarized light*.

In 1828, the British physicist William Nicol (1768?–1851) produced a device that took advantage of the different directions in which these plane-polarized light rays traveled inside the crystal of Iceland spar. He began with a rhombohedral crystal of the substance (one with every face a parallelogram) and cut it diago-

Nicol prism

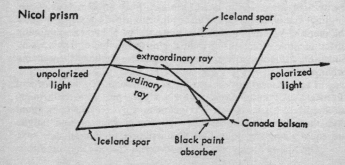

nally. The two halves were cemented together again by way of a layer of Canada balsam (a resin from a tree called the balsam fir). Light entering the crystal would be split up into two plane-polarized rays traveling in slightly different directions. One ray would strike the Canada balsam at an angle such that total reflection would take place. The reflected ray would then strike a painted section of the prism and be absorbed. The other ray, striking the Canada balsam at a slightly different angle, would be transmitted, pass into the other half of the crystal and out into the open air again.

The light emerging from such a *Nicol prism,* then, would consist of a single plane-polarized ray, representing about half the original light intensity.

Suppose the light passing through a Nicol prism is made to pass through a second Nicol prism. If the second prism is aligned in the same fashion as the first, the light will pass through the second unhindered. (That is like a rope with up-down waves passing through two picket fences, one behind the other. Neither fence gets in the way.)

But suppose the second Nicol prism is rotated through a small angle. The polarized light emerging from the first prism cannot get through the second prism in full intensity. There is some loss (as there would be in the up-down waves of the rope if the slats of wood in the second picket fence were tipped a little into the diagonal).

The amount of light that would get through the second prism would decrease as the angle through which that prism was rotated increased. Once the second prism was rotated through 90°, no light at all could get through.

The second prism can thus be used to determine the exact plane in which the light issuing from the first prism is polarized. By twisting the second prism and noticing the alignment at which the light one sees through it is at maximum brightness, one finds the plane of polarization. If one sees no light at all, the alignment of the second prism is at right angles to the plane of polarization. Since it is difficult to judge exactly where maximum or minimum brightness is, the second prism may be manufactured in such a way as to consist of two prisms set at a slight angle to each other. If one is aligned properly, the other is slightly off. Looking through an eyepiece, then, one would see one half distinctly brighter than the other. By adjusting the alignment so that both halves are equally bright, one locates the plane of polarization.

The first prism in such an instrument, the one that produces the polarized light, is the *polarizer*. The second, which determines the plane of polarization, is the *analyzer*. The instrument as a whole is a *polariscope*.

Even before the Nicol prism was invented, it was discovered by the French physicist Jean Baptiste Biot (1774–1862), in 1815, that polarized light, traveling through solutions of certain substances, or through certain transparent crystals, would have its plane of polarization shifted.

Suppose, for instance, that between the two prisms of a polariscope is a cylindrical vessel containing air and that the prisms are aligned in the same direction. If water is poured into the tube, nothing happens; the two halves of the field as seen in the eyepiece remain equally bright. The plane of polarization of the light has not been altered by passing through water. If instead of pure water a sugar solution had been placed in the tube, the two halves seen in the eyepiece would become unequally bright. The analyzer would have to be turned through some definite angle to make them equally bright again. That angle would represent the amount by which the plane of polarized light had been rotated by the sugar solution.

The size of this angle depends on various factors: the concentration of the solution and the nature of the substance dissolved; the distance through which light travels within the solution; the wavelength of the light; and the temperature of the solution. If one standardizes these factors, and either observes or calculates what the angle of rotation would be for light of a wavelength equal to that produced by a sodium vapor lamp, traveling through one decimeter of a solution containing 1 gram per cubic centimeter at a temperature of 20°C, then one obtains the *specific rotation*.

The value of the specific rotation is characteristic for every transparent system. For many this value is 0°—that is, the plane of polarized light is not rotated at all. Such systems are *optically inactive*. Systems that do rotate the plane of polarized light are *optically active*.

Some optically active systems rotate the plane of polarized light in a clockwise direction. This is taken as a right-handed turn, and such systems are *dextrorotatory*. Others turn the light in a counterclockwise direction and are *levorotatory*.

In 1848, the French chemist Louis Pasteur (1822–1895) was able to show that the optical activity of transparent crystals was dependent on the fact that such crystals were asymmetric. Further, if such asymmetric crystals could be fashioned into two mirror-image forms, one would be dextrorotatory, and the other, levorotatory. The fact that certain solutions also exhibited optical activity argued that asymmetry must be present in the very molecules of these substances. In 1874, the Dutch physical chemist Jacobus Hendricus van't Hoff (1852–1911) presented a theory of molecular structure that accounted for such asymmetry in optically active substances. A discussion of this, however, belongs more properly in a book on chemistry, and I will go no further into the subject here.

Nicol prisms are not the only means by which beams of plane-polarized light can be formed. There are some types of crystal that do not merely split the light into two plane-polarized beams but absorb one and transmit the other. Crystals of iodo-quinine sulfate will do this. Unfortunately, it is impossible to manufacture large useful crystals of this material since such crystals are fragile and disintegrate at the least disturbance.

In the mid-1930's, however, it occurred to a Harvard undergraduate, Edwin Herbert Land (1909–   ) that single large crystals were not necessary. Tiny crystals, all oriented in the same direction, would serve the purpose. To keep them so oriented, and to keep them from further disintegration, they could be embedded in a sheet of transparent, flexible plastic. Land quit school in 1936 to go into business and produced what is now known as *Polaroid*. It can serve all the functions of Nicol prisms more economically and conveniently (though not with quite the same precision).

As Malus had discovered, beams of polarized light can also be produced by reflection, at some appropriate angle, from material such as glass; the exact size of the angle depends on the index of refraction of the material. "Sunglasses" made of Polaroid can block most of this reflected polarized light and cut down glare.

Thus, the nineteenth century saw light established not merely as a wave form but as a transverse wave form; if this solved many problems, it also raised a few.

6

The Ether

Absolute Motion

If light is a wave form, then it seemed to most scientists, up to the beginning of the twentieth century, that something must be waving. In the case of water waves, for instance, water molecules move up and down; in the case of sound waves, the atoms or molecules of the transmitting medium move to and fro. Something, it seemed, would therefore have to be present in a vacuum; something that would move either up and down, or to and fro, in order to produce light waves.

This something, whatever it was, did not interfere with the motions of the heavenly bodies in any detectable way, so it seemed reasonable to suppose it to be nothing more than an extremely rarefied gas. This extremely rarefied gas (or whatever it was that filled the vacuum of space) was called *ether*, from a word first used by Aristotle to describe the substance making up the heavens and the heavenly bodies (see page 1–6). Ether might also be the substance through which the force of gravitation was transmitted, and this might be identical with the ether that did or didn't transmit light. In order to specify the particular ether that transmitted light (in case more than one variety existed) the phrase *luminiferous ether* ("light-carrying ether") grew popular in the nineteenth century.

In connection with the ether, the difference in properties

between transverse and longitudinal waves becomes important. Longitudinal waves can be conducted by material in any state: solid, liquid or gaseous. Transverse waves, however, can only be conducted through solids, or, in a gravitational field, along liquid surfaces (see page I–158). Transverse waves cannot be conducted through the body of a liquid or gas. It was for this reason that early proponents of the wave theory of light, assuming the ether to be a gas, also assumed light to consist of longitudinal waves that could pass through a gas rather than transverse waves that could not.

When the question of polarization, however, seemed to establish the fact that light consisted of transverse waves, the concept of the ether had to be drastically revised. The ether had to be a solid to carry transverse light waves; it had to be a substance in which all parts were fixed firmly in place.

If that were so, then if a portion of the ether were distorted at right angles to the motion of a light beam (as seemed to be required if light were a transverse wave phenomena), the forces tending to hold that portion in place would snap it back. That portion would overshoot the mark, snap back from the other direction, overshoot the mark again, and so on. (This is what happens in the case of water waves, where gravity supplies the force necessary for snap-back, and in sound waves, where intermolecular forces do the job.)

The up and down movement of the ether thus forms the light wave. Moreover, the rate at which a transverse wave travels through a medium depends on the size of the force that snaps back the distorted region. The greater the force, the faster the snap-back, the more rapid the progression of the wave. With light traveling at over 186,000 miles per second, the snap-back must be rapid indeed, and the force holding each portion of the ether in place was calculated to be considerably stronger than steel.

The luminiferous ether, therefore, must at one and the same time be an extremely tenuous gas, and possess a rigidity greater than that of steel. Such a combination of properties is hard to visualize,* but during the mid-nineteenth century, physicists la-

---

* Such a combination is against "common sense," but this must never be allowed to stand in the way of the acceptance of a hypothesis. We experience only a very limited portion of the universe and are sensitive to only a very limited range of phenomena. It is therefore dangerous to suppose that what seems familiar to us is and must be true of all the universe in all its aspects. Thus, it is only "common sense" to suppose that the earth is flat and motionless, and this argument was strenuously used to oppose the notion that the earth was spherical and in motion.

bored hard to work out the consequences of such a rigid-gas and to establish its existence. They did this for two reasons. First, there seemed no alternative, if light consisted of transverse waves. Secondly, the ether was needed as a reference point against which to measure motion. This second reason is extremely important, for without such a reference point, the very idea of motion becomes vague, and all of the nineteenth century development of physics becomes shaky.

To explain why that should be, let us suppose that you are on a train capable of moving at a uniform velocity along a perfectly straight set of rails with vibrationless motion. Ordinarily, you could tell whether your train were actually in motion by the presence of vibration, or by inertial effects when the train speeds up, slows down, or rounds a curve. However, with the train moving uniformly and vibrationlessly, all this is eliminated and the ordinary methods for noting that you are in motion are useless.

Now imagine that there is one window in the train through which you can see another train on the next track. There is a window in that other train, and someone is looking out at you through it. Speaking to you by sign language, he asks, "Is my train moving?" You look at it, see clearly that it is motionless, and answer, "No, it is standing still." So he gets out and is killed at once, for it turns out that both trains are moving in the same direction at 70 miles per hour with respect to the earth's surface.

Since both trains were moving in the same direction at the same speed, they did not change position with respect to each other, and each seemed motionless to an observer on the other. If there had been a window on the other side of each train, one could have looked out at the scenery and noted it moving rapidly toward the rear of the train. Since we automatically assume that the scenery does not move, the obvious conclusion would be that the train is actually in motion even though it does not seem to be.

Again, suppose that in observing the other train, you noted that it was moving backward at two miles an hour. You signal this information to the man in the other train. He signals back a violent negative. He is standing still, he insists, but you are moving forward at two miles an hour. Which one of you is correct?

To decide that, check on the scenery. It may then turn out that Train A is motionless while Train B is actually moving backward at two miles an hour. Or Train B may be motionless while Train A is moving forward at two miles an hour. Or Train A may be moving forward at one mile an hour while Train B is moving backward at one mile an hour. Or both trains may be moving forward: Train A at 70 miles an hour and Train B at 68 miles an hour. There are an infinite number of possible motions, with respect to the earth's surface, that can give rise to the observed motion of Train A and Train B relative to each other.

Through long custom, people on trains tend to downgrade the importance of the relative motion of one train to another. They consider it is the motion with respect to the earth's surface that is the "real" motion.

But is it? Suppose a person on a train, speeding smoothly along a straight section of track at 70 miles an hour, drops a coin. He sees the coin fall in a straight line to the floor of the train. A person standing by the wayside, watching the train pass and able to watch the coin as it falls, would see that it was subjected to two kinds of motion. It falls downward at an accelerating velocity because of gravitational force and it shares in the forward motion of the train, too. The net effect of the two motions is to cause the coin to move in a parabola (see page I–39).

We conclude that the coin moves in a straight line relative to the train and in a parabola relative to the earth's surface. Now which is the "real" motion? The parabola? The person on the train who is dropping the coin may be ready to believe that although he seems to himself to be standing still, he is "really" moving at a velocity of 70 miles an hour. He may not be equally ready to believe that a coin that he sees moving in a straight line is "really" moving in a parabola.

This is a very important point in the philosophy of science. Newton's first law of motion (page I–24) states that an object not subjected to external forces will move in a straight line at constant speed. However, what seems a straight line to one observer does not necessarily seem a straight line to another observer. In that case, what meaning does Newton's first law have? What is straight-line motion, anyway?

Throughout ancient and medieval times, almost all scholars believed that the earth was affixed to the center of the universe and never budged from that point. The earth, then, was truly motionless. It was (so it was believed) in a state of *absolute rest*. All motion could be measured relative to such a point at absolute rest, and then we would have *absolute motion*. This absolute motion would be the "true" motion upon which all observers could agree. Any observed motion that was not equivalent to the absolute motion was the result of the absolute motion of the observer.

There was some question, of course, as to whether the earth

were truly motionless, even in ancient times. The stars seemed to be moving around the earth in 24 hours at a constant speed. Was the earth standing still and the celestial sphere turning, or was the celestial sphere standing still and the earth turning? The problem was like that of the two trains moving relative to each other, with the "real motion" unverifiable until one turned to look at the scenery. In the case of the earth and the celestial sphere, there was no scenery to turn to and no quick decision, therefore, upon which everyone could agree.

Most people decided it was the celestial sphere that turned, because it was easier to believe that than to believe that the vast earth was turning without our being able to feel that we were moving. (We still speak of the sun, moon, planets and stars as "rising" and "setting.") In modern times, however, for a variety of reasons better discussed in a book on astronomy, it has become far more convenient to suppose that the earth is rotating rather than standing still.

In such a case, while the earth as a whole is not at absolute rest, the axis may be. However, by the beginning of modern times, more and more astronomers were coming to believe that even the earth's axis was not motionless. The earth, all of it, circled madly about the sun along with the other planets. No part of it was any more at rest than was any train careening along its surface. The train might have a fixed motion relative to the earth's surface, but that was not the train's "true" motion.

For a couple of centuries after the motion of the earth had come to be accepted, there was still some excuse to believe that the sun might be the center of the universe. The sun visibly rotated, for sunspots on its surface circled its globe in a steady period of about 27 days. However, the sun's axis might still represent that sought-for state of absolute rest.

Unfortunately, it became clearer and clearer, as the nineteenth century approached, that the sun was but a star among stars, and that it was moving among the stars. In fact, we now know that just as the earth moves around the sun in the period of one year, the sun moves about the center of our Galaxy in a period of 200,000,000 years. And, of course, the Galaxy itself is but a galaxy among galaxies and must be moving relative to the others.

By mid-nineteenth century, there was strong reason to suppose that no material object anywhere in the universe represented a state of absolute rest, and that absolute motion could not therefore be measured relative to any material object. This might have raised a serious heart-chilling doubt as to the universal validity of Newton's laws of motion, on which all of nineteenth century physics· was based. However, a material object was not needed to establish absolute motion.

It seemed to nineteenth century physicists that if space were filled with ether, it was fair to suppose that this ether served only to transmit forces such as gravity and waves such as those of light, and was not itself affected, overall, by forces. In that case, it could not be set into motion. It might vibrate back and forth, as in transmitting light waves, but it would not have an overall motion. The ether, then, might be considered as being at absolute rest. All motion became absolute motion if measured relative to the ether. This ether-filled space, identical to all observers, aloof, unchanging, unmoving, crossed by bodies and forces without being affected by them, a passive container for matter and energy, is *absolute space*.

In Newton's time and for two centuries afterward, there was no way of actually measuring the motion of any material body relative to the ether. Nevertheless, that didn't matter. In principle, absolute motion was taken to exist, whether it was practical to measure it or not, and the laws of motion were assumed to hold for such absolute motion and, therefore, must surely hold for all relative motions (which were merely one absolute motion added to another absolute motion).

The Michelson-Morley Experiment

In the 1880's, however, it appeared to Michelson (the latter-day measurer of the velocity of light) that a method of determining absolute motion could be worked out.

Light consists of waves of ether, according to the view of the time, and if the ether moved, it should carry its own vibrations (light) with it. If the ether were moving away from us, it would carry light away from us and therefore delay light in its motion toward us—reduce the velocity of light, in other words. If the ether were moving away from us at half the velocity of light, then light would lose half its velocity relative to ourselves and therefore take twice as long to get to us from some fixed point. Similarly, if the ether were moving toward us, light would reach us more quickly than otherwise.

To be sure, physicists were assuming that the ether itself was not moving under any circumstances. However, the earth must, it seemed, inevitably be moving relative to the ether. In that case, ıf the earth is taken as motionless, then the ether would seem to be moving relative to us, fixed as we are to the earth. There would seem to be what came to be called an "ether wind."

If there were no ether wind at all, if the earth were at absolute rest, then light would travel at the same velocity in all directions. To be sure, it actually seems to do just this, but surely that is only because the ether wind is moving at a very small velocity compared to the velocity of light; therefore, light undergoes only minute percentage changes in its velocity with shift in direction. In view of the difficulty of measuring light's velocity with any accuracy in the first place, it would not be surprising that small differences in velocity with shifting direction would go unnoticed.

Michelson, however, in 1881, invented a device that was perhaps delicate enough to do the job.

In this device, light of a particular wavelength falls upon a glass plate at an angle of 45°. The rear surface of the glass plate is "half-silvered." That is, the surface has been coated with enough silver to reflect half the light and allow the remaining half to be transmitted. The transmitted light emerges, traveling in the same direction it had been traveling in originally, while the reflected light moves off at right angles to that direction. Both light beams are reflected by a mirror and travel back to the half-silvered glass plate. Some of the originally reflected beam now passes through, while some of the originally transmitted beam is now reflected. In this way, the two beams join again.*

In effect a single beam of light has been split in two; the two halves have been sent in directions at right angles to each other, have returned, and have been made to join in a combined beam again.

The two beams, joining, set up interference fringes, as did the two beams in Young's experiment. One of the mirrors can be adjusted so that the length of the journey of the beam of light to that particular mirror and back can be varied. As the mirror is adjusted, the interference fringes move. From the number of

---

* The transmitted light travels through the glass plate once on its way outward. When returning it strikes the silver and is reflected without entering the glass. The reflected light travels through the glass plate once in reaching the silver, travels through a second time after reflection, and a third time on its return trip For this reason, a second glass plate, identical with the first, is placed in the path of the transmitted light. It must travel through it both going and returning, and now each beam of light has traveled through the same total thickness of glass. Michelson had to be very careful to make sure that both beams of light received identical treatment in every possible way.

fringes that pass the line of sight when the mirror is moved a certain distance, the wavelength of the light can be determined. The greater the number of fringes passing the line of sight, the shorter the wavelength.

Michelson determined wavelengths of light with his instrument, which he called the *interferometer* ("to measure by interference"), so precisely that he suggested the wavelength of some particular spectral line be established as the fundamental unit of length. At the time, this fundamental unit had just been established as the *International Prototype Meter*. This was the distance between two fine marks on a bar of platinum-iridium alloy kept at Sèvres, a suburb of Paris.

In 1960, Michelson's suggestion was finally accepted and the fundamental unit of length became a natural phenomenon rather than a man-made object. The orange-red spectral line of a variety of the rare gas krypton was taken as the standard. The meter is now set officially equal to 1,650,763.73 wavelengths of this light.

But Michelson was after bigger game than the determination of the wavelengths of spectral lines. He considered the fact that the beam of light in the interferometer was split into two halves that traveled at right angles to each other. Suppose one of these two light rays happened to be going with the ether wind. Its velocity would be c (the velocity of light with respect to the ether) plus v (the velocity of the light source with respect to the

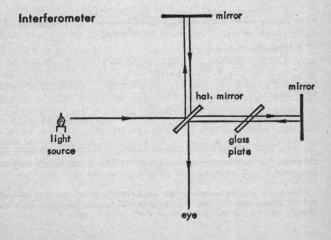

Interferometer

ether). If the distance of the reflecting mirror from the half-silvered prism is taken as d, then the time it would take the light to pass from the half-silvered prism to the reflecting mirror would be $d/(c+v)$. After reflection, the light would move over the distance d in precisely the opposite direction. Now it would be moving into the ether wind, and it would be slowed down, its overall velocity being $c-v$. The time taken for its return would be $d/(c-v)$. The total time (t_1) taken by that beam of light to go and return is therefore:

$$t_1 = \frac{d}{c+v} + \frac{d}{c-v} = \frac{2dc}{c^2-v^2}$$ (Equation 6–1)

Meanwhile, however, the second half of the beam is going at right angles to the first; it also returns at right angles to the first. It is going neither with the ether wind nor against it. It is going "crosswind" both ways.

The time taken by the light beam to go and return crosswind (t_2) can be calculated with the help of plane geometry* and turns out to be:

$$t_2 = \frac{2d}{\sqrt{c^2-v^2}}$$ (Equation 6–2)

If we divide Equation 6–1 by Equation 6–2, we will determine the ratio of the time taken to cover the ground with-and-against the ether wind and the time taken to cover the same distance crosswind. We would have:

$$\frac{t_1}{t_2} = \frac{2dc}{c^2-v^2} \div \frac{2d}{\sqrt{c^2-v^2}} = \frac{c\sqrt{c^2-v^2}}{c^2-v^2}$$ (Equation 6–3)

The expression at the extreme right of Equation 6–3 is of the form, $a\sqrt{x}/x$, and if both numerator and denominator of such an expression is divided by $\sqrt{x}$, the equivalent expression $a/\sqrt{x}$ is obtained. Equation 6–3 can therefore be simplified to:

$$\frac{t_1}{t_2} = \frac{c}{\sqrt{c^2-v^2}}$$ (Equation 6–4)

Further simplification can be obtained if both numerator and denominator are multipled by $\sqrt{1/c^2}$. (The multiplication of the numerator and denominator of a fraction by the same quantity

* Anyone curious concerning the details of this calculation may try pages 807–810 of my book *The New Intelligent Man's Guide to Science*, Basic Books, 1965.

does not, of course, alter the value of the expression as a whole.)

The numerator of Equation 6–4 then becomes $c\sqrt{1/c^2}$ or c/c or 1. The denominator becomes $\sqrt{c^2 - v^2}\ \sqrt{1/c^2}$ or $\sqrt{c^2(1/c^2) - v^2(1/c^2)}$ or $\sqrt{1 - v^2/c^2}$. Equation 6–4 can therefore be expressed as:

$$\frac{t_1}{t_2} = \frac{1}{\sqrt{1 - \dfrac{v^2}{c^2}}}$$

(Equation 6–5)

If the light source is at rest with respect to the ether, $v = 0$ and $t_1/t_2 = 1$. In that case the time taken by the beam of light going with-and-against the ether wind is the same as the time taken by the beam of light to go crosswind. (Indeed, the time is the same for light beams going in any direction.) If the movable mirror is adjusted so that the two beams of light travel exactly the same distance, they will return exactly in step and there will be no interference fringes. Furthermore there will be no interference fringes if the instrument is then turned so as to have the light beams travel in changed directions.

However, if the light source is moving with respect to the ether, then v is greater than 0, $1 - v^2/c^2$ is less than 1, and t_1/t_2 is greater than 1. The light traveling with-and-against the ether would then take longer to cover a fixed distance than the light traveling crosswind would. To be sure, the ratio is not very much greater than 1 for any reasonable velocity relative to the ether. Even if the light source were moving at one-tenth the velocity of light (so that v was equal to the tremendous figure of 30,000 kilometers per second), the ratio would be only 1.005. At ordinary velocities, the ratio would be very small indeed.

Nevertheless, the difference in time would be enough to throw the wavelengths of the two beams of light out of step and set up interference fringes. Naturally, you could not know in advance which direction would be with-and-against the ether wind and which would be crosswind, but that would not matter. The instrument could be pointed in some direction at random, and the movable mirror could be adjusted so as to remove the interference fringes. If the instrument were turned now, the light beams would change direction and be affected differently by the ether wind so that interference fringes would appear.

From the spacing of the fringes one could determine the velocity of the light source relative to the ether. Since the light source was firmly attached to the earth, this was equivalent to finding the velocity of the earth relative to the ether——that is, the absolute motion of the earth. Once that was done, all bodies, as long as their motions relative to the earth were known, would have absolute motions that were known.

Michelson obtained the help of an American chemist, Edward Williams Morley (1838–1923), and in 1886 he tried this experiment. Michelson had tried it alone, before, but never under conditions that he found satisfactory. Now he and Morley dug down to bedrock to anchor the interferometer, and balanced the instrument with fantastic precautions against error.

Over and over again, they repeated the experiment and always the results were the same——negative! Once they adjusted the device to remove interference fringes, those fringes did not show up to any significant extent when the interferometer was reoriented. One might have thought that they just happened to be unlucky enough to try the experiment at a time when the earth happened to be motionless with respect to the ether. However, the earth travels in an ellipse about the sun and changes the direction of its motion every moment. If it were at rest with respect to the ether on one day, it could not be at rest the next.

Michelson and Morley made thousands of observations over many months, and in July, 1887, finally announced their conclusion. There was no ether wind!

I have gone into detail concerning this experiment because of the shocking nature of the result. To say there was no ether wind meant there was very likely no way of determining absolute motion. In fact, the very concept of absolute motion suddenly seemed to have no meaning. And if that was so, what would become of Newton's laws of motion and of the whole picture of the universe as based upon those laws?

Physicists would have been relieved to find that the *Michelson-Morley experiment* was wrong and that there was an ether wind after all. However, the experiment has been repeated over and over again since 1887. In 1960, devices far more accurate than even the interferometer were used for the purpose, and the result was always the same. There is no ether wind. This fact simply had to be accepted, and the view of the universe changed accordingly.

The FitzGerald Contraction

Naturally, attempts were made to explain the results of the Michelson-Morley experiment in terms of the ether. The most successful attempt was that of the Irish physicist George Francis FitzGerald (1851–1901), who in 1893 proposed that all objects grew shorter in the direction of their absolute motion, being shortened, so to speak, by the pressure of the ether wind. Distances between two bodies moving in unison would likewise shorten in the direction of the motion, since the two bodies would be pushed together by the ether wind. The amount of this "foreshortening" would increase with the velocity of the absolute motion, of course, as the pressure of the ether wind rose.

FitzGerald suggested that at any given velocity, the length (L) of an object or of the distance between objects would have a fixed ratio to the length (L_o) of that same object or distance at rest; and L_o may be termed the *rest-length*. This ratio would be expressed by the quantity $\sqrt{1 - v^2/c^2}$, where c is the velocity of light in a vacuum, and v is the velocity of the body, both relative to the ether. In other words:

$$L = L_o\sqrt{1 - v^2/c^2} \qquad \text{(Equation 6–6)}$$

The FitzGerald ratio is equal to the denominator of the expression in Equation 6–5, which represents the ratio of the distances traveled by the two beams of light in the interferometer. Mutiplied by FitzGerald's ratio, the value in Equation 6–5 becomes 1. The distance covered by the beam of light moving with-and-against the ether wind is now decreased by foreshortening to just exactly the extent that would allow the beam to cover the distance in the same time as was required by the beam traveling crosswind. In other words, the existence of the ether wind would make one of the beams take a longer time, but the existence of the *FitzGerald contraction* produced by the same ether wind allows the beam to complete its journey in the same time as one would expect if there were no ether wind.

The two effects of the ether wind cancel perfectly, and this has reminded a thousand physicists of a passage from a poem in Lewis Carroll's *Through the Looking-Glass*.

> "But I was thinking of a plan
>   To dye one's whisker's green,
> And always use so large a fan
>   That they could not be seen."

Carroll's book was written in 1872, so it could not have deliberately referred to the FitzGerald contraction, but the reference is a perfect one, just the same.

The contraction is extremely small at ordinary velocities. The earth moves in its orbit about the sun at 30 kilometers per second (relative to the sun), which by earthly standards is a great velocity. If v is set equal to 30 and this is inserted into the Fitz-Gerald ratio, we have $\sqrt{1 - (30)^2/(300,000)^2}$, which is equal to 0.999995. The foreshortened diameter of the earth in the direction of its motion would then be 0.999995 of its diameter perpendicular to that direction (assuming the earth to be a perfect sphere). The amount of the foreshortening would be 62.5 meters.

If the earth's diameter could be measured in all directions, and the direction in which the diameter were abnormally short could be located, then the direction of the earth's motion relative to the ether could be determined. Furthermore, from the size of the abnormal decrease in diameter, the absolute velocity of the earth relative to the ether could be worked out.

But there is a difficulty. This difficulty lies not in the smallness of the foreshortening, because no matter how small it is all might be well if it could be detected in principle. It cannot be detected, however, as long as we remain on earth. While on earth, all the instruments we could conceivably use to measure the earth's diameter would share in the earth's motion and in its foreshortening. The foreshortened diameter would be measured with foreshortened instruments, and no foreshortening would be detected.

We could do better if we could get off the earth and, without sharing in the earth's motion, measure its diameter in all directions (very accurately) as it speeds past. This is not exactly practical but it is something that is conceivable in principle.

To make such a thing practical, we must find something that moves very rapidly and in whose motion we do not ourselves share. Such objects would seem to be the speeding subatomic particles* that have motions relative to the surface of the earth of anywhere from 10,000 kilometers per second to nearly the speed of light.

The FitzGerald contraction becomes very significant at such super-velocities. The velocity might be high enough, for instance, for the length of the moving body to be foreshortened to only half its rest-length. In that case $\sqrt{1 - v^2/c^2} = 1/2$, and if we solve for v, we find that it equals $\sqrt{3c^2/4}$. Since $c = 300,000$ kilometers per second, $\sqrt{3c^2/4} = 260,000$ kilometers per second. At this ferocious velocity, seven-eighths that of light, an object is foreshortened to half its rest-length, and some subatomic particles move more rapidly (relative to the earth's surface) than this.

---

* These particles, much smaller than atoms, make up the atomic structure. They will be discussed in some detail in the third volume of this book.

At still more rapid velocities, foreshortening becomes even more marked. Suppose the velocity of a body becomes equal to the velocity of light. Under those conditions v is equal to c, and FitzGerald's ratio becomes $\sqrt{1 - c^2/c^2}$, which equals 0. This means that by Equation 6–6 the length of the moving body (L) is equal to its rest-length (L_o) multiplied by zero. In other words, at the velocity of light, all bodies, whatever their length at rest, have foreshortened completely and have become pancakes of ultimate thinness.

But then what happens if the velocity of light is exceeded? In that case, v becomes greater than c, the expression v^2/c^2 becomes greater than 1, and the expression $1 - v^2/c^2$ becomes a negative number. FitzGerald's ratio is the square root of a negative number, and this is what mathematicians call an "imaginary number." A length represented by an imaginary number has mathematical interest, but no one has been able to work out the physical meaning of such a length.

This was the first indication that the velocity of light might have some important general significance in the universe—as something that might, in some fashion, represent a maximum velocity. To be sure, no subatomic particle has ever been observed to move at a velocity greater than that of light in a vacuum, although velocities of better than 0.99 times that of light in a vacuum have been observed. At such velocities, the subatomic particles ought to be wafer-thin in the direction of their motion, but alas, they are so small that it is completely impractical to try to measure their length as they speed past, and one cannot tell whether they are foreshortened or not. If, however, the length of the speeding subatomic particles will not do as a practical test of the validity of the FitzGerald contraction, another property will. . . .

The FitzGerald contraction was put into neat mathematical form, and extended, by the Dutch physicist Hendrik Antoon Lorentz (1853–1928) so that the phenomenon is sometimes referred to as the *Lorentz-FitzGerald contraction*.

Lorentz went on to show that if the FitzGerald contraction is applied to subatomic particles carrying an electric charge, one could deduce that the mass of a body must increase with motion in just the same proportion as its length decreases. In short, if its mass while moving is m and its *rest-mass* is m_o, then:

$$m = \frac{m_o}{\sqrt{1 - v^2/c^2}}$$

(Equation 6–7)

Again, the gain in mass is very small at ordinary velocities. At a velocity of 260,000 kilometers per second, the mass of the moving body is twice the rest-mass, and above that velocity it increases ever more rapidly. When the velocity of a moving body is equal to that of light, $v = c$ and Equation 6–7 becomes $m = m_o/0$. This means that the mass of the moving body becomes larger than any value that can be assigned to it. (This is usually expressed by saying that the mass of the moving body becomes infinite.) Once again, velocities greater than light produced masses expressed by imaginary numbers, for which there seems no physical interpretation. The key importance of the velocity of light in a vacuum is again emphasized.

But the very rapidly moving charged subatomic particles possessing velocities up to 0.99 times that of light increase markedly in mass; and whereas the length of speeding subatomic particles cannot be measured as they fly past, their mass can be measured easily.

The mass of such particles can be obtained by measuring their inertia—that is, the force required to impose a given acceleration upon them. In fact, it is this quantity of inertia that Newton used as a definition of mass in his second law of motion (see page I–30.)

Charged particles can be made to curve in a magnetic field. This is an acceleration imposed upon them by the magnetic force, and the radius of curvature is the measure of the inertia of the particle and therefore of its mass.

From the curvature of the path of a particle moving at low velocity, one can calculate the mass of the particle and then predict what curvature it will undergo when it passes through the same magnetic field at higher velocities, provided its mass remains constant. Actual measurement of the curvatures for particles moving at higher velocities showed that such curvatures were less marked than was expected. Furthermore, the higher the velocity, the more the actual curvature fell short of what was expected. This could be interpreted as an increase in mass with velocity, and when this was done the relationship followed the Lorentz equation exactly.

The fan had slipped and the green whiskers could be seen. The Lorentz equation fit the observed facts. Since it was based on the FitzGerald equation, the phenomenon of foreshortening also fit the facts, and this explained the negative results of the Michelson-Morley experiment.

CHAPTER 7

Relativity

The Special Theory

If the gain in mass of a speeding charged particle is the result of its motion relative to the ether, then a new method of measuring such motion might offer itself. Suppose some charged particles are measured as they speed along in one direction, others as they speed in another direction, and so on. If all directions are taken into account, some particles are bound to be moving with the ether wind, while others, speeding in the opposite direction, are moving against it. Those moving against the ether wind (one might suspect) will have a more rapid motion relative to the ether and will gain more mass than will those moving at the same velocity (relative to ourselves) with the ether wind. By the changes in gain of mass as direction is changed, the velocity of the ether wind, and therefore the absolute motion of the earth, can be determined.

However, this method also fails, exactly as the Michelson-Morley experiment failed. The gain in mass with motion is the same no matter in which direction the particles move. What's more, all experiments designed to measure absolute motion have failed.

In 1905, in fact, a young German-born, Swiss physicist, Albert Einstein (1879–1955), had already decided that it might be useless to search for methods of determining absolute motion. Suppose that one took the bull by the horns and simply decided that it was impossible to measure absolute motion by any conceivable method* and considered the consequences.

That, then, was Einstein's first assumption: that all motion must be considered relative to some object or some system of objects arbitrarily taken as being at rest; and that any object or system of objects (any *frame of reference*, that is) can be taken, with equal validity, as being at rest. There is no object, in other words, that is more "really" at rest than any other.

Since in this view all motion is taken as relative motion only, Einstein was advancing what came to be called the *theory of relativity*. In his first paper on the subject, in 1905, Einstein considered only the special case of motion at constant velocity; therefore, this portion of his views is his *special theory of relativity*.

Einstein then made a second assumption: that the velocity of light in a vacuum, as measured, would always turn out to be the same, whatever the motion of the light source relative to the observer. (Notice that I speak of the velocity "as measured.")

This measured constancy of the velocity of light seems to be in violation of the "facts" about motion that had been accepted since the days of Galileo and Newton.

Suppose a person throws a ball past us and we measure the horizontal velocity of the ball relative to ourselves as x feet per minute. If the person is on a platform that is moving in the opposite direction at y feet per minute and throws the ball with the same force, its horizontal velocity relative to ourselves ought to be $x - y$ feet per minute. If the platform were moving in the same direction he threw the ball, the horizontal velocity of the ball relative to ourselves ought to be $x + y$ feet per minute.

This actually seems to be the situation as observed and measured in real life. Ought it not therefore be the same if the person were "throwing" light out of a flashlight instead of a ball out of his fist?

In order to make Einstein's second assumption hold true,

---

* This is not the same as saying flatly that there is no absolute motion. All that scientists know of the physical universe is based, directly or indirectly, on observation and measurement. If there is some phenomenon that can neither be observed nor measured under any conceivable circumstances, then, as far as the world of experimental science is concerned, it can be treated as though it does not exist. Whether it "really" exists, though it can't be either observed or measured, even in principle, is a question that may amuse philosophers and theologians but is completely irrelevant to scientists.

we must suppose that this situation does not hold for light at all and, in fact, that it does not hold for the ball either.

Suppose that the effect of the moving platform on the speed of the ball is not quite as great as we suspect and that when the motion of the platform is added to that of the ball, the overall velocity of the ball is a vanishingly small amount smaller than $x + y$. Again, when the motion of the platform is subtracted from that of the ball the overall velocity of the ball is a vanishingly small amount greater than $x - y$. Suppose, too, that this discrepancy increases as x and y increase, but that at all velocities of material bodies, which we were capable of observing before 1900, the discrepancy remained far too small to measure. Consequently, it would be very natural for us to come to the conclusion that the velocity was exactly $x + y$ or exactly $x - y$, and that this would remain true for all speeds.

But if one could observe very great velocities, velocities of the order of thousands of kilometers per second, the discrepancy would become great enough to notice. If one added the velocity y to the velocity x, the combined velocity would then be noticeably less than $x + y$ and might be hardly any greater than x alone. Similarly, if y were subtracted from x, the combined velocity might be considerably larger than $x - y$ and hardly less than x alone. Finally, at the speed of light, the effect of the movement of the source of the moving body will have declined to zero so that $x + y = x$ and $x - y = x$, regardless of how great y is. And that is another way of expressing Einstein's second assumption.

In order to save that assumption, in fact, it is necessary to add velocities in such a way that the sum never exceeds the velocity of light. Suppose, for instance, a platform is moving forward (with respect to ourselves) at 290,000 kilometers a second, or only 10,000 kilometers a second less than the velocity of light in a vacuum. Suppose, further, that a ball is thrown forward from the platform at a velocity, relative to the platform, of 290,000 kilometers a second. The velocity of the ball relative to ourselves ought to be $290,000 + 290,000$ kilometers a second in that forward direction, but at those velocities the effect of the moving platform has so decreased that the overall velocity is, in point of fact, only 295,000 kilometers per second and is still less than the velocity of light.

Indeed, this can be expressed mathematically. If two velocities (V_1 and V_2) are added, then the new velocity (V) according to Newton would be $V = V_1 + V_2$. According to Einstein, the new velocity would be:

$$V = \frac{V_1 + V_2}{1 + \dfrac{V_1 V_2}{C^2}}$$

Where C is equal to the velocity of light in a vacuum. If V_1 is equal to C, then Einstein's equation would become:

$$V = \frac{C + V_2}{1 + \dfrac{CV_2}{C^2}} = (C + V_2)\left(\frac{C}{C + V_2}\right) = C$$

In other words, if one velocity were equal to the speed of light, adding another velocity to it, even again up to the speed of light, would leave the total velocity merely at the speed of light.

To put it briefly, it is possible to deduce from Einstein's assumption of the constant measured velocity of light that the velocity of any moving body will always be measured as less than the velocity of light.*

It seems strange and uncomfortable to accept so unusual a set of circumstances just to save Einstein's assumption of the measured constancy of the velocity of light. Nevertheless, whenever it has been possible to measure the velocity of light, that velocity has always been placed at one constant value, and whenever it has been possible to measure the velocity of speeding bodies, that velocity has always been less than the velocity of light. In short, no physicist has yet detected any phenomenon that can be taken as violating either Einstein's assumption of relativity of motion or his assumption of measured constancy of light; and they have looked assiduously, you may be sure.

Einstein could also deduce from his assumptions the existence of the Lorentz-FitzGerald contraction as well as the Lorentz gain of mass with motion. Furthermore, he showed that it was not only electrically charged particles that gained mass with motion, but uncharged particles as well. In fact, all objects gained mass with motion.

It might seem that there is scarcely any reason to crow over the special theory so far. What is the difference between starting with the assumption of the Lorentz-FitzGerald contraction and deducing from it the measured constancy of the velocity of light,

---

* This is often expressed as "a body cannot move faster than light" but that is not quite right. It is only the *measured velocity* that is less than the measured velocity of light. It is quite conceivable that there are objects in the universe that are traveling at velocities (relative to ourselves) that are greater than the velocity of light, but we could not see such bodies or sense them in any way and therefore could not measure their velocities.

or starting with the assumption of the measured constancy of the velocity of light and deducing from it the Lorentz-FitzGerald contraction?

If that were all, there would be no significant difference, indeed. However, Einstein combined his assumption concerning the measured constancy of the velocity of light with his first assumption that all motion is relative.

This meant that foreshortening or mass-gain was not a "real" phenomenon but only a change in measurement. The amount by which length was decreased or mass increased was not something that could be absolutely determined but differed from observer to observer.

To consider what this means, imagine two identical spaceships moving in opposite directions in a non-collision course, each spaceship possessing equipment that will enable it to measure the length and mass of the other spaceship as it passes by.

Spaceship X watches Spaceship Y flash by (in a particular direction) at 260,000 kilometers per second, and at this velocity Spaceship Y is measured as being only half its rest-length and fully twice its rest-mass. Spaceship X, which to the people on board seems to be motionless, is to them, naturally, exactly at rest-length and rest-mass.

But the people on Spaceship Y have no sensation of moving (any more than we have the sensation of speeding through space on our voyage around the sun). The people on Spaceship Y feel themselves to be motionless and find themselves to be at rest-length and rest-mass. What they see is Spaceship X flashing by (in the opposite direction) at 260,000 kilometers per second. To them it is Spaceship X that is measured as being only half its rest-length and fully twice its rest-mass.

If the observers could communicate while in motion, they could have a glorious argument. Each could say, "I am at rest and you are moving. I am normal length, you are foreshortened. I am normal mass, you are heavy."

Well, which one is really "right"?

The answer is neither and both. It is not a question, you see, of what has "really" happened to length and mass or of which ship is "really" foreshortened or over-massive. The question is only of measurement. (It is—to make a trivial analogy—like measuring the side of a rectangle that is four meters by two meters and then arguing about whether the length of the rectangle is "really" four meters or "really" two meters. It depends on the side you are measuring.)

But suppose you attempt to perform some kind of test that will, perhaps, reach beyond the measurement to the "reality." Suppose, for instance, you brought the two ships together and compared them directly to see which was shorter and heavier. This cannot actually be done within the bounds of Einstein's special theory since that deals only with uniform motion. To bring the ships together means that at least one of them must turn round and come back and thus undergo non-uniform, or accelerated, motion. Even if we did this, however, and imagined the two ships side by side and at rest relative to each other, after having passed each other at super-velocities, we could make no decision as to "realities." Being at rest with respect to each other, each would measure the other as being normal in length and mass. If there had been a "real" change in length or mass in either ship in the past, there would remain no record of that change.

Despite everything, it is difficult to stop worrying about "reality." It is heartening, then, to remember that there have been times when we have abandoned a spurious "reality" and have not only survived but have been immeasurably the better for it.

Thus, a child is pretty certain he knows what "up" and "down" is. His head points "up," his feet point "down" (if he is standing in the normal fashion); he jumps "up," he falls "down." Furthermore, he discovers soon enough that everyone around him agrees as to which direction is "up" and which "down."

If a child with such convictions is shown a picture of the earth's globe, with the United States above and Australia below, and with little Americans standing head-up and little Australians standing head-down, his first thought may well be, "But that's impossible. The little Australians would fall off."

Of course, once the effect of the gravitational force is understood (and this was understood as long ago as Aristotle, at least as far as the earth itself was concerned; see page I–5ff.) then there is no longer fear that anyone would fall off any part of the earth. However, you might still be questioning the nature of "up" and "down." You might call up an Australian on long-distance telephone and say, "I am standing head-up, so you must be standing head-down." He would reply, "No, no. I am clearly standing head-up, so it must be you who is standing head-down."

Do you see, then, how meaningless it is to ask now who is right and who is "really" standing head-up? Both are right and both are wrong. Each is standing head-up in his own frame of reference, and each is standing head-down in the other's frame of reference.

Most people are so used to this that they no longer see a "relative up" and a "relative down" as being in violation of "common sense." In fact, it is the concept of the "absolute up" and the "absolute down" that seems a violation now. If anyone seriously argued that the Australians walked about suspended by their feet, he would be laughed at for his ignorance.

Once the tenets of the relativistic universe are accepted (at as early an age as possible) it, too, ceases to go against common sense.

Mass-Energy Equivalence

During the nineteenth century, chemists were increasingly convinced that mass could neither be created nor destroyed (the *law of conservation of mass*). To Lorentz and Einstein, however, mass was created as velocity increased, and it was destroyed as velocity decreased. To be sure, the changes in mass are vanishingly small at all ordinary velocities, but they are there. Where, then, does created mass come from, and where does destroyed mass go?

Let's begin by considering a body of a given mass (m) subjected to a given force (f). Under such conditions the body undergoes an acceleration (a), and from Newton's second law of motion (see page I-30) one can state that $a = f/m$. The presence of an acceleration means that the velocity of the body is increasing, but in the old Newtonian view this did not affect the mass of the body, which remained constant. If the force is also viewed as remaining constant, then f/m was constant and a, the acceleration, was also constant. As a result of such a constant acceleration, the velocity of the body (in the Newtonian view) would increase indefinitely and would reach any value you care to name—if you wait long enough.

In the Einsteinian universe, however, an observer measuring the velocity of an object under a continuing constant force can never observe it to exceed the velocity of light in a vacuum. Consequently, though its velocity increases under the influence of a constant force, that velocity increases more and more slowly, and as the velocity approaches that of light, it increases exceedingly slowly. In short, the acceleration of a body under the influence of a constant force decreases as the velocity increases and becomes zero when the velocity reaches that of light.

But, again from Newton's second law of motion, the mass of a body is equal to the force exerted upon it divided by the acceleration produced by that force—that is, $m = f/a$. If the force is constant and the acceleration decreases with velocity, then a decreases with velocity while f does not; consequently, f/a increases with velocity. And this means, since $m = f/a$, that mass increases with velocity. (Thus, the increase of mass with velocity can be deduced from Einstein's assumption of the measured constancy of the velocity of light in a vacuum.)

When a body is subjected to a force, it gains kinetic energy, which is equal to one half its mass times the square of its velocity ($e_k = 1/2mv^2$; see page I–95). In the Newtonian view this increase in kinetic energy results only from the increase in velocity, for mass is considered unchanging. In the Einsteinian view the increase in kinetic energy is the result of an increase in both velocity and mass.

Where mass is not involved in energy changes (as in the Newtonian view) it is natural to think of mass as something apart from energy and to think that, on the one hand, there is a law of conservation of energy, and on the other, a law of conservation of mass, and that the two are independent.

Where mass changes and is thus intimately involved in energy changes (as in the Einsteinian view), it is natural to think of mass and energy as different aspects of the same thing, so a law of conservation of energy would include mass. (To make that perfectly clear, in view of our previous convictions, we sometimes speak of the *law of conservation of mass-energy*, but the word "mass" is not really needed.)

Motion does not create mass in any real sense; mass is merely one aspect of a general increase in kinetic energy gained from the force that is maintained by the expenditure of energy elsewhere in the system.

But now suppose the law of conservation of energy (including mass) remains valid in the relativistic universe (and so far it seems to have done so). According to this law, although energy can be neither created nor destroyed, it can be changed from one form to another. This would seem to mean that a certain quantity of mass could be converted into a certain quantity of other forms of energy such as heat, for instance; and that a certain quantity of a form of energy such as heat might, conceivably, be converted into a certain quantity of mass. And this, indeed, Einstein insisted upon.

This equivalence of mass and energy announced by Einstein in his 1905 paper was of great use to the physicists of the time. The discovery of radioactivity nine years earlier (something I will discuss in volume III) had revealed a situation in which energy seemed to be created endlessly out of nowhere. Once the special theory of relativity pointed the way, scientists searched for disappearing mass and found it.

It may seem surprising that no one noticed the interchange of mass and energy until Einstein pointed it out theoretically. The reason for that rests with the nature of the equivalence—in the determination of exactly how much energy is equivalent to how much mass.

To determine that, let's consider the reciprocal of the FitzGerald ratio, which is $1/\sqrt{1 - v^2/c^2}$. This can also be written, according to algebraic convention, as $(1 - v^2/c^2)^{-1/2}$. An expression written in this last fashion can be said to belong to a family of the type $(1 - b)^{-a}$. By the binomial theorem (a mathematical relationship first worked out by Newton himself), the expression $(1 - b)^{-a}$ can be expanded into an endless series of terms that begins as follows: $1 + ab + 1/2(a^2 + a)\ b^2 + \ldots$.

To apply this expansion to the reciprocal of the FitzGerald ratio, we must set $a = 1/2$ and $b = v^2/c^2$. The FitzGerald ratio then becomes: $1 + v^2/2c^2 + 3v^4/8c^4 \ldots$.

Since c, the velocity of light, may be considered to have a constant value, the second and third terms (and, indeed, all the subsequent terms of this infinite series) grow larger as v increases. But v reaches a maximum when the velocity of a moving body attains the velocity of light (at least, we can measure no higher velocity). Therefore the various terms are then at their maximum value, and at $v = c$ the series becomes $1 + 1/2 + 3/8. \ldots$ As you see, the second term is, at most, less than the first, while the third term is, at most, less than the second, and so on.

The decrease is even sharper at lower velocities, and successive terms become rapidly more and more insignificant. When $v = c/2$ (150,000 kilometers per second), the series is $1 + 1/8 + 3/128. \ldots$ When $v = c/4$ (75,000 kilometers per second), the series is $1 + 1/32 + 3/2048. \ldots$

In a decreasing series of this sort it is possible to show that the tail end of the series (even though it includes an infinite number of terms) reaches a finite and small total. We can therefore eliminate all but the first few terms of the series, and consider those first few a good approximation of the whole series.

At ordinary velocities, for instance, all the terms of the series except the first (which is always 1) become such small fractions that they can be ignored completely. In that case, the reciprocal of the FitzGerald ratio can be considered as equal to 1 with a high decree of accuracy (which is why changes in length and mass with motion went unnoticed until the twentieth century). To make it still more accurate, especially at very high velocities, we can include the first two terms of the series. That is accurate enough for all reasonable purposes, and we need not worry at all about the third term or any beyond it.

We can say, then, with sufficient accuracy that:

$$\frac{1}{\sqrt{1 - v^2/c^2}} = 1 + v^2/2c^2 \qquad \text{(Equation 7–1)}$$

Now let us return to the Lorentz mass relationship (Equation 6–7), which states that the mass (m_1) of a body in motion is equal to its rest-mass (m_0) divided by the FitzGerald ratio. This is equivalent to saying that m_1 is equal to m_0 multiplied by the reciprocal of the FitzGerald ratio; therefore, using the new expression for that reciprocal given in Equation 7–1, we can write the mass relationship as follows:

$$m_1 = m_0(1 + v^2/2c^2) \qquad \text{(Equation 7–2)}$$
$$= m_0 + m_0 v^2/2c^2$$

The increase in mass as a result of motion is $m_1 - m_0$, and we can call this difference simply m. If we solve Equation 7–2 for $m_1 - m_0$ (that is, for m), we find that:

$$m = m_0 v^2/2c^2 = \tfrac{1}{2}m_0 v^2/c^2 \qquad \text{(Equation 7–3)}$$

The expression $\tfrac{1}{2}m_0 v^2$, found in the right-hand portion of Equation 7–3, happens to be the value of the kinetic energy of the moving body (kinetic energy is equal to $\tfrac{1}{2}mv^2$, see page I–95), if it possesses its rest-mass. Actually, it possesses a slightly higher mass due to its motion, but except for extremely high velocities, the actual mass is only very slightly higher than the rest-mass—so little higher in fact that we can let $\tfrac{1}{2}m_0 v^2$ equal its kinetic energy and be confident of a high degree of accuracy. If we let this kinetic energy be represented as e, then Equation 7–3 becomes:

$$m = e/c^2 \qquad \text{(Equation 7–4)}$$

Remember that m represents the gain of mass with motion. Since very rapid motion, representing a very high value of e, (the kinetic energy) produces only a very small increase in mass, we see quite plainly that a great deal of ordinary energy is equivalent to a tiny quantity of mass. Equation 7–4, which by clearing fractions can be written as the much more familiar:

$$e = mc^2 \qquad \text{(Equation 7–5)}$$

can be used to calculate this equivalence.

In the cgs system (see page I–32), where all the units are in terms of centimeters, grams and seconds, the value of c (the velocity of light in a vacuum) is 30,000,000,000 centimeters per second. The value of c^2 is therefore 900,000,000,000,000,000,000 cm²/sec². If we set the value of m at 1 gram, then mc^2 is equal to 900,000,000,000,000,000,000 gm-cm²/sec²; or, since 1 gm-cm²/sec² is defined as an "erg," 1 gram of mass is equal to 900,-000,000,000,000,000,000, ergs of energy.

One kilocalorie is equal to 41,860,000,000 ergs. This means that 1 gram of mass is equivalent to 21,500,000,000 kilocalories. The combustion of a gallon of gasoline liberates about 32,000 kilocalories. The mass equivalence of this amount of energy is 32,000/21,500,000,000 or 1/670,000 of a gram. This means that in the combustion of a full gallon of gasoline, the evolution of energy in the form of heat, light, the mechanical motion of pistons, and so on, involves the total loss to the system of 1/670,000 of a gram of mass. It is small wonder that chemists and physicists did not notice such small mass changes until they were told to look for it.

On the other hand, if whole grams of mass could be converted wholesale into energy, the vast concentration of energy produced would have tremendous effects. In Volume III the steps by which it was learned how to do this will be outlined. The results are the *nuclear bombs* that now threaten all mankind with destruction and the *nuclear reactors* that offer it new hope for the future.

Furthermore, Equation 7–5 offered the first satisfactory explanation of the source of energy of the sun and other stars. In order for the sun to radiate the vast energies it does, it must lose 4,600,000 tons of mass each second. This is a vast quantity by human standards but is insignificant to the sun. At this rate it can continue to radiate in essentially unchanged fashion for billions of years.

The *Einstein equation*, $e = mc^2$, as you see, is derived entirely from the assumption of the constant measured velocity of light, and the mere existence of nuclear bombs is fearful evidence of the validity of the special theory of relativity. It is no wonder that of all equations in physics, $e = mc^2$ has most nearly become a household word among the general population of non-physicists.

Relative Time

Einstein deduced a further conclusion from his assumptions and went beyond the Lorentz-FitzGerald dealings with length and mass to take up the question of time as well.

The passing of time is invariably measured through some steady periodic motion: the turning of the earth, the dripping of water, the beating of a pendulum, the oscillation of a pendulum, even the vibration of an atom within a molecule. However, the changes in length and mass with increasing velocity, must inevitably result in a slowing of the period of all periodic motion. Time therefore must be measured as proceeding more slowly as velocity relative to the observer is increased.

Again, the FitzGerald ratio is involved. That is, the time lapse (t) observed on a body moving at a given velocity relative to the time lapse at rest (t_0) is as follows:

$$t = t_0 \sqrt{1 - v^2/c^2}$$

(Equation 7–6)

At a velocity of 260,000 kilometers per second past an observer, t would equal $t_0/2$. In other words, it would take one hour of the observer's time for half an hour to seem to pass on the moving object. Thus, if an observer's clock said 1:00 and a clock on the moving object also said 1:00, then one hour later the observer's clock would say 2:00, but the clock on the moving object would say only 1:30.

At a velocity equal to that of light, t would equal 0. It would then take forever for the clock on the moving object to show any time lapse at all to the observer. As far as the observer would be able to note, the clock on the moving object would always read 1:00; time would stand still on the object. This slowing of time with motion is called *time dilatation*.

Strange as this state of affairs seems, it has been checked in the case of certain short-lived subatomic particles. When moving slowly, they break down in a certain fixed time. When moving very rapidly, they endure considerably longer before decaying. The natural conclusion is that we observe a slowing of time for the speedily moving particles. They still decay in, say, a millionth of a second, but for us that millionth of a second seems to stretch out because of the rapid motion of the particle.

As in the case of length and mass, this change in time is a change in measurement only (as long as the conditions of the special theory are adhered to), and this varies with the observer.

Suppose, for instance, we return to Spaceship X and Spaceship Y as they flash past each other. The men on Spaceship X, watching Spaceship Y flash by at 260,000 kilometers a second and observing a pendulum clock on board Spaceship Y, would see the clock beat out its seconds at two-second intervals. Everything on Spaceship Y would take twice the usual time to transpire (or so it would seem to the observer on Spaceship X). The very atoms would move at only half their usual speed.

The people on Spaceship Y would be unaware of this, of course. Considering themselves at rest, they would insist that it was Spaceship X that was experiencing slowed time. (Indeed, if the spaceships had flashed by each other in such a way that each measured the velocity of the other as equaling the velocity of light, each would insist that it had observed time on the other ship having come to a full halt.)

This question of time is trickier, however, than that of length and mass. If the spaceships are brought together after the flash-by, and placed at mutual rest, length and mass are now "normal" and no record is left of any previous changes, so there need be no worries about "reality."

But, as for time, consider . . . Once at mutual rest, the clocks are again proceeding at the "normal" rate and time lapses are equal on both ships. Yet, previous changes in time lapses *have* left a record. If one clock has been slowed and has, in the past, registered only half an hour while the other clock was registering an hour, the first clock would now be half an hour slow! Each ship would firmly claim that the clock on the other ship had been registering time at a slower-than-normal rate, and each would expect the other's clock to be slow.

Would this be so? Would either clock be slow? And if so, which?

This is the *clock paradox*, which has become famous among physicists.

There is no clock paradox if the conditions of the special theory are adhered to strictly—that is, if both ships continue eternally in uniform motion. In that case, they can never be brought together again, and the difference in measurement remains one that can never be checked against "reality."

In order to bring the ships together, at least one of them must slow down, execute a turn, speed up and overtake the other.

In all this it undergoes non-uniform velocity, or acceleration, and we are promptly outside the special theory.

Einstein worked on problems of this sort for ten years after having enunciated his special theory and, in 1915, published his *general theory of relativity*, in which the consequences of non-uniform or accelerated motion are taken up. This is a much more subtle and difficult aspect of relativity than is the special theory, and not all theoretical physicists entirely agree on the consequences of the general theory.

Suppose we consider our spaceships as alone in the universe. Spaceship Y executes the slowdown, turn, and speedup that brings it side by side with Spaceship X. But, by the principle of relativity, the men on Spaceship Y have every right to consider themselves at rest. If they consider themselves at rest, then it is Spaceship X that (so it seems to them) slows down, turns, and then backs up to them. Whatever effect the men on Spaceship X observe on Spaceship Y, the men on board Spaceship Y will observe on Spaceship X. Thus, it might be that when the two ships are finally side by side, the two clocks will somehow tell the same time.

Actually, though, this will not occur, for the two spaceships are not alone in the universe. The universe is filled with a vast amount of matter, and the presence of this amount of matter spoils the symmetry of the situation of Spaceships X and Y.

Thus, if Spaceship Y executes its turn, Spaceship X observes it make that turn. But as Spaceship X considers itself at rest, it continues to see the rest of the universe (the stars and galaxies) slip past it at a constant, uniform velocity reflecting its own constant, uniform velocity. In other words, Spaceship X sees *only* Spaceship Y and nothing else undergo non-uniform motion.

On the other hand if Spaceship Y considers itself at rest, it observes that not only does Spaceship X seem to undergo an acceleration but also all the rest of the universe.

To put it another way, Spaceship Y and Spaceship X both undergo non-uniform motion relative to each other, but the universe as a whole undergoes non-uniform motion only relative to Spaceship Y. The two ships, naturally enough, are influenced differently by this tremendous difference in their histories, and when they are brought together, it is Spaceship Y (which has undergone non-uniform motion relative to the universe as a whole) that carries the slowed clock. There is no paradox here, for the crews on both ships must have observed the non-uniform motion of the universe relative to Spaceship Y, and so both agree on the difference in histories and cannot seek refuge in a "my-frame-of-reference-is-as-good-as-yours" argument.

Now suppose a space traveler leaves earth and, after a while, is traveling away from us at a speed nearly that of light. If we could observe him as he traveled, we would see his time pass at only perhaps one-hundredth the rate ours does. If he observed us, he would see our time pass at only one-hundredth the rate his does. If the space traveler wanted to return, however, he would have to turn, and he would experience non-uniform motion relative to the universe as a whole. In other words, in turning, he would observe the entire universe turning about him, if he insisted on considering himself at rest. The effect of this is to make the time lapse less for him as far as both he and the stay-at-home earthmen are involved.

The round trip may have seemed to him to have lasted only a year, but on the earth a hundred years would have passed. If the space traveler had a twin brother, left behind on earth, that brother would long since have died of old age, while the traveler himself would scarcely have aged. (This is called the *twin paradox*.) It is important, however, to realize that the space traveler has not discovered a fountain of youth. He may have aged only a year in an earth-century, but he would only have lived a year in that earth-century. Moreover, no matter what his velocity, time would never appear, either to him or to the earth-bound observers, to go backward. He would never grow younger.

The variation of the rate at which times passes as velocity changes destroys our concept of absoluteness of time. Because of this, it becomes impossible to locate an event in time in such a way that all observers can agree. In addition, no event can be located in time until some evidence of the event reaches the observer, and that evidence can only travel at the velocity of light.

As a simple example, consider the space traveler returning to earth, after having experienced a time lapse of one year, and finding that his twin brother had died fifty years earlier by earth time. To the traveler this may seem impossible, since fifty years earlier (to him) his twin brother had not even been born yet.

In fact, in the mathematical treatment of the theory of relativity, it does not make sense to deal with space alone or time alone. Rather the equations deal with a fusion of the two (usually called space-time). To locate a point in space-time, one must express a value for each of three spatial dimensions, plus a value for time; time being treated somewhat (but not exactly) like the ordinary three dimensions. It is in that sense that time is sometimes spoken of as the "fourth dimension."

It is sometimes argued that the existence of relative time makes it possible to measure a velocity of more than that of light. Suppose, for instance, a spaceship travels from earth to a planet ten light-years distant and does this at so great a velocity that time dilatation makes it seem to the crewmen that only one year has passed in the course of the voyage.

Since the ship has traveled in one year a distance traversed by light in ten years, has not the ship traveled at ten times the velocity of light?

The answer is it has not. If the crewmen were to argue that they had, they would be measuring the time lapse of one year against their own frame of reference, and the distance of the planet from earth (ten light-years) by earth's frame of reference. They must ask instead: What is the distance of the destination-planet from earth in the frame of reference of the ship?

In the ship's frame of reference, the ship is, of course, motionless, while the universe, including earth and the destination-planet, slips backward past it at an enormous velocity. The entire universe is foreshortened, as one would expect from the FitzGerald contraction (see page 97), and the distance from earth to the destination-planet is much less than ten light-years. The distance is less than one light-year, in fact, so the ship can traverse that distance in one year without having exceeded the velocity of light.

Again, although the ship took only one year to get to its destination, this did not mean they beat light there, even though a light beam, released from earth simultaneously with the ship, would have taken ten years to cover a ten-light-year distance. That ten-year time lapse would be true only in the earth's frame of reference. For the light beam's own frame of reference, since it travels at the velocity of light, the rate of passage of time would decline to zero, and the light beam would get to Alpha Centauri (or to any spot in the universe, however distant) in no time at all.

Nor can one use this to argue that in the light beam's own frame of reference its velocity is then infinite; for in the light beam's own frame of reference, the total thickness of the universe, in the direction of its travel, is foreshortened to zero, and of course it would take no time for light to cross a zero-thickness universe even if its velocity is the finite one of 300,000 kilometers per second.

The General Theory

One of the basic assumptions in the special theory was that it was impossible to measure absolute motion; that any observer had the privilege of considering himself at rest; and that all frames of reference were equally valid.

Yet when we consider non-uniform motion (outside the realm of special theory) the possibility arises that this is not so.

Suppose, for instance, that two spaceships are moving side by side at uniform velocity. The crewmen on each ship can consider both themselves and the other ship to be at rest. Then, suddenly, Spaceship Y begins to move forward with reference to Spaceship X.

The crewmen on Spaceship X could maintain that they were still at rest while Spaceship Y had begun to move forward at an accelerating velocity. The crewmen on Spaceship Y, however, could maintain that, on the contrary, they were at rest while Spaceship X had begun to move backward at an accelerating velocity. Is there any way, now, to decide between these conflicting observations?

In the case of such non-uniform motion, perhaps. Thus, if Spaceship Y were "really" accelerating forward, the men within it would feel an inertial pressure backward (as you are pressed back into your seat when you step on your car's gas-pedal.) On the other hand, if Spaceship X is accelerating backward, the men within it would feel an inertial pressure forward (as you lurch toward the windshield when you step on your car's brake). Consequently, the crews of the spaceships could decide which ship was "really" moving by taking note of which set of crewmen felt inertial pressures.

From this, one could perhaps determine absolute motion from the nature and size of inertial effects. Einstein, in his general theory of relativity, worked out what properties the universe must possess to prevent the determination of absolute motion in the case of non-uniform motion.

The Newtonian view of mass had dealt, really, with two kinds of mass. By Newton's second law of motion, mass was defined through the inertia associated with a body. This is "inertial mass." Mass may also be defined by the strength of the gravitational field to which it gives rise. This is "gravitational mass." Ever since Newton, it had been supposed that the two masses were really completely identical, but there had seemed no way of proving it. Einstein did not try to prove it; he merely assumed that inertial mass and gravitational mass were identical and went on from there.

It was then possible to argue that both gravitation and inertial effects were not the property of individual bodies alone, but of the interaction of the mass of those bodies with all the remaining mass in the universe.

If a spaceship begins to accelerate in a forward direction, the crewmen feel an inertial pressure impelling them to the rear. But suppose the crewmen in the spaceship insist on regarding themselves as at rest. They must then interpret their observations of the universe as indicating that all the stars and galaxies outside the ship are moving backward at an accelerating velocity. The accelerating motion backward of the distant bodies of the universe drags the crewmen back, too, producing an inertial effect upon them, exactly as would have happened if the universe had been considered at rest and the ship as accelerating forward.

In short, inertial effects cannot be used to prove that the ship is "really" accelerating. The same effect would be observed if the ship were at rest and the universe were accelerating. Only *relative* non-uniform motion is demonstrated by such inertial effects: either a non-uniform motion of the ship with reference to the universe or a non-uniform motion of the universe with reference to the ship. There is no way of demonstrating which of these two alternatives is the "real" one.

We might also ask if the earth is "really" rotating. Through most of man's history, the earth was assumed motionless because it seemed motionless. After much intellectual travail, its rotation was demonstrated to the satisfaction of scientists generally and to those non-scientists who followed the arguments or were willing to accept the word of authority. But is it "really" rotating?

One argument in favor of the rotation of the earth rests on the existence of the planet's equatorial bulge. This is explained as the result of a centrifugal effect that must surely arise from a rotation. If the earth did not rotate, there would be no centrifugal effect and it would not bulge. The existence of the bulge, therefore, is often taken as proof of a "real" rotation of the earth.

This argument might hold, perhaps, if the earth were alone in the universe, but it is not. If the earth is considered motionless, for argument's sake, one must also think of the enormous mass of the universe revolving rapidly about the earth. The effect of this enormous revolving mass is to pull out the earth's equatorial bulge—just as the centrifugal effect would if the earth rotated and the rest of the universe were motionless. One could always explain all effects of rotation equally well in either frame of reference.

You might also argue that if the earth were motionless and the rest of the universe revolved about it, the distant stars, in order to travel completely around their gigantic orbits about earth in a mere 24 hours, must move at many, many times the velocity of light. From this one might conclude that the rotation of the universe about the earth is impossible and that therefore the earth is "really" rotating. However, if the universe is considered as rotating about the earth and if the distant stars are traveling, in consequence, at great velocities, the FitzGerald contraction will reduce the distances they must cover to the point where their velocity will be measured as less than that of light.

Of course, one might raise the argument that it simply isn't reasonable to suppose the entire universe is revolving about the earth—that one must naturally prefer to believe that it is the earth's rotation that produces the apparent revolution of the universe. Similarly it is much more sensible to believe that a spaceship is accelerating forward rather than to suppose an entire universe is accelerating backward past one motionless ship.

This is true enough, and it is so much more reasonable to assume a rotating earth (or a moving ship) that astronomers will continue to assume it, regardless of the tenets of relativity. However, the theory of relativity does not argue that one frame of reference may not be simpler or more useful than another—merely that one frame of reference is not more *valid* than another.

Consider that at times it is the motionlessness of the earth that is assumed because that makes for greater simplicity. A pitcher throwing a baseball never takes into account the fact that the earth is rotating. Since he, the ball, and the waiting batter are all sharing whatever velocity the earth possesses, it is easier for the pitcher to assume that the earth is motionless and to judge the force and direction of his throw on that basis. For him the motionless-earth frame of reference is more useful than the rotating-earth frame of reference—yet that does not make the motionless-earth frame of reference more valid.

Gravitation

In his general theory, Einstein also took a new look at gravitation. To Newton it had seemed that if the earth revolved about the sun there must be a force of mutual attraction between the earth and the sun. Einstein showed that one could explain the revolution of the earth about the sun in terms of the geometry of space.

Consider an analogy. A putter is addressing a golf ball toward the cup over a level green. The golf ball strikes the edge of the cup and dips in. It is, however, going too fast so that it spins about the vertical side of the cup (bobsled fashion) and emerges at the other end, rolling in a new direction. It has partly circled the center of the cup, yet no one would suppose that there was a force of attraction between the golf ball and the center of the cup.

Let us imagine a perfectly level, frictionless, putting green of infinite extent. A ball struck by the golf club will continue on forever in a perfectly straight line.

But what if the putting green is uneven; if there are bumps and hollows in it? A ball rising partly up the side of a bump will curve off in a direction away from the center of the bump. A ball dropping down the side of a hollow will curve toward the center of the hollow. If the bumps and hollows are, for some reason, invisible and undetectable, we might be puzzled at the occasional deviations of the balls from straight-line motion. We might suppose the existence of hidden forces of attraction or repulsion pulling or pushing the ball this way and that.

Suppose one imagined a cone-shaped hollow with steep sides on such a green. A ball can be visualized as taking up a closed "orbit" circling around and around the sides like a bobsled speeding endlessly along a circular bank. If friction existed, the circling ball would lose kinetic energy and, little by little, sink to the bottom of the cone. In the absence of friction, it would maintain its orbit.

It is not difficult to form an analogous picture of the Einsteinian version of gravity. Space-time would be a four-dimensional analogy of a flat putting green, if it were empty of matter. Matter, however, produces "hollows"; the more massive the matter, the deeper the "hollow." The earth moves about the sun as though it were circling the sun's hollow. If there were friction in space, it would slowly sink to the bottom of the "hollow" (that is, spiral into the sun). Without friction, it maintains its orbit indefinitely. The elliptical orbit of the earth indicates that the orbit about the "hollow" is not perfectly level with the flatness of the four-dimensional putting green. (The orbit would be a circle, if it were.) A slight tilt of the orbit produces a slight ellipticity, while a more marked tilt produces greater ellipticity.

It is these "hollows" produced by the presence of matter that give rise to the notion of *curved space*.

The consequences of the special theory of relativity—mass increase with motion and the equivalence of mass and energy, for instance—were easily demonstrated. The validity of the general theory was much more difficult to prove. Einstein's picture of gravitation produces results so nearly like those of Newton's picture that it is tempting to consider the two equivalent and then accept the one that is simpler and more "common sense," and that, of course, is the Newtonian picture.

However, there remained some areas where the consequences of the Einsteinian picture were indeed somewhat different from those of the Newtonian picture. By studying those consequences, one might choose between the two on some basis more satisfying than that of mere simplicity. The first such area involved the planet Mercury.

The various bodies in the solar system move, in the Newtonian view, in response to the gravitational forces to which they are subjected. Each body is subjected to the gravitational forces of every other body in the universe, so the exact and complete solution of the motions of any body is not to be expected. However, within the solar system, the effect of the gravitational field of the sun is overwhelming. While the gravitational fields of a few other bodies quite close to the body whose motion is being analyzed are also significant, they are minor. If these are taken into account, the motion of a planet of the solar system can be explained with a degree of accuracy that satisfies everybody. If residual disagreements between the predicted motion and the actual motion remain, the assumption is that some gravitational effect has been ignored.

The presence of a discrepancy in the motion of Uranus, for instance, led to a search for an ignored gravitational effect, and the discovery, in the mid-nineteenth century, of the planet Neptune.

At the time of Neptune's discovery, a discrepancy in the motion of Mercury, the planet nearest the sun, was also being studied. Like the other planets, Mercury travels in an ellipse about the sun, with the sun at one of the foci of the ellipse. This means that the planet is not always at the same distance from the sun. There is a spot in its orbit where it is closest to the sun, the *perihelion*, and a spot at the opposite end of the orbit where it is farthest from the sun, the *aphelion*. The line connecting the two is the *major axis*. Mercury does not repeat its orbit exactly, but moves in such a way that the orbit is actually a rosette, with the major axis of the ellipse slowly revolving.

This can be explained by the gravitational effect of nearby planets on Mercury, but it cannot all be explained. After all known gravitational effects are accounted for, the actual rate at which the major axis (and its two extreme points, the perihelion and aphelion) turned was slightly greater than it ought to have been—greater by 43.03 seconds of arc per century. This meant that the major axis of Mercury's orbit made a complete turn, and an unexplained one, in 3,000,000 years.

Leverrier, one of the men who had discovered Neptune, suggested that an undiscovered planet might exist between Mercury and the sun, and that the gravitational effect of this planet on Mercury could account for that additional motion of the perihelion. However, the planet was never found, and even if it existed (or if a belt of planetoids of equivalent mass existed near the sun) there then would also be gravitational effects on Venus, and these have never been detected.

The situation remained puzzling for some seventy years until Einstein in 1915 showed that the general theory of relativity altered the view of gravity by just enough to introduce an additional factor that would account for the unexplained portion of the motion of Mercury's perihelion. (There would be similar but much smaller effects on the planets farther from the sun—too small to detect with certainty.)

Einstein also predicted that light beams would be affected by gravity, a point that was not allowed for in the Newtonian view. The light of stars passing very close to the sun, for instance, would be affected by the geometry of space and would bend inward toward the center of the sun. Our eyes would follow the ray of light backward along the new direction and would see the star located farther from the center of the sun than it really was. The effect was very small. Even if light just grazed the sun, the shift in a star's position would be only 1.75 seconds of arc, and if the light passed farther from the sun, the shift in the star's position would be even less.

Of course, the light of stars near the sun cannot ordinarily be observed. For a few minutes during the course of a total eclipse, however, they can be. At the time the general theory was published, World War I was in progress and nothing could be done. In 1919, however, the war was over and a total eclipse was to be visible from the island of Principe in the Gulf of Guinea off West Africa. Under British auspices an elaborate expedition was sent to the island for the specific purpose of testing the general theory. The positions of the stars in the neighborhood of the sun were measured and compared with their positions a half year later when the sun was in the opposite end of the sky. The results confirmed the general theory.

Finally, Einstein's theory predicted that light would lose energy if it rose against gravity and would gain energy if it "fell," just as an ordinary object would. In the case of a moving object such as a ball, this loss of energy would be reflected as a loss of velocity. However, light could only move at one velocity; therefore the loss of energy would have to be reflected in a declining frequency and increasing wavelength. Thus, light leaving a star would undergo a slight "red shift" as it lost energy. The effect was so small, however, that it could not be measured.

However, stars had just been discovered (*white dwarfs*) which were incredibly dense and which had gravitational fields thousands of times as intense as those of ordinary stars. Light leaving such a star should lose enough energy to show a pronounced red shift of its spectral lines. In 1925, the American astronomer Walter Sydney Adams (1876–1956) was able to take the spectrum of the white dwarf companion of the star Sirius, and to confirm this prediction.

The general theory of relativity had thus won three victories in three contests over the old view of gravitation. All, however, were astronomical victories. It was not until 1960 that the general theory was brought into the laboratory.

The key to this laboratory demonstration was discovered in 1958 by the German physicist Rudolf Ludwig Mössbauer (1929–), who showed that under certain conditions a crystal could be made to produce a beam of gamma rays* of identical wavelengths. Gamma rays of such wavelengths can be absorbed by a crystal similar to that which produced it. If the gamma rays are of even slighly different wavelength, they will not be absorbed. This is called the *Mössbauer effect*.

Now, then, if such a beam of gamma rays is emitted downward so as to "fall" with gravity, it gains energy and its wavelength becomes shorter—if the general theory of relativity is correct. In falling just a few hundred feet, it should gain enough energy for the decrease in wavelength of the gamma rays, though very minute, to become sufficiently large to prevent the crystal from absorbing the beam.

Furthermore, if the crystal emitting the gamma ray is moved upward while the emission is proceeding, the wavelength of the

---

* Gamma rays are a form of light-like radiation that will be discussed in Volume III of this book.

gamma ray is increased through the Doppler-Fizeau effect. The velocity at which the crystal is moved upward can be adjusted so as to just neutralize the effect of gravitation on the falling gamma ray. The gamma ray will then be absorbed by the absorbing crystal. Experiments conducted in 1960 corroborated the general theory of relativity with great accuracy, and this was the most impressive demonstration of its validity yet.

It is not surprising, then, that the relativistic view of the universe is now generally accepted (at least until further notice) among physicists of the world.

CHAPTER 8

Quanta

Black-Body Radiation

The theory of relativity does not flatly state that an ether does not exist. It does, however, remove the need for one, and if it is not needed, why bother with it?

Thus, the ether was not needed to serve as an absolute standard for motion since relativity began by assuming that such an absolute standard did not exist and went on to demonstrate that it was not needed. Again, the ether is not needed as a medium to transmit the force of gravity and prevent "action at a distance." If gravity is a matter of the geometry of space-time and is not a transmitted force, the possibility of action at a distance does not arise.

This still leaves one possible use for the ether—that of serving as a medium for transmitting light waves across a vacuum. A second paper written by Einstein in 1905 (in addition to his paper on special relativity) wiped out that possibility, too. Einstein's work on relativity had evolved out of the paradox concerning light that was turned up by the Michelson-Morley experiment (see page 99ff.). Einstein's second paper arose out of a different paradox, also concerning light, that had also arisen in the last decades of the nineteenth century. (It was for this second paper that he later received the Nobel Prize.)

This second paradox began with Kirchhoff's work on spectroscopy (see page 58). He showed that a substance that absorbed certain frequencies of light better than others would also emit those frequencies better than others once it was heated to incandescence.

Supppose, then, one imagined a substance capable of absorbing all the light, of all frequencies, that fell upon it. Such a body would reflect no light of any frequency and would therefore appear perfectly black. It is natural to call such a substance a *black body* for that reason. If a black body is brought to incandescence, its emission should then be as perfect as its absorption, by Kirchhoff's rule. It should emit light in all frequencies, since it absorbs in all frequencies. Furthermore, since it absorbs light at each frequency more efficiently than a non-black body would, it must radiate more efficiently at each frequency, too.

Kirchhoff's work served to increase the interest of physicists in the quantitative aspects of radiation, and in the manner in which such radiation varied with temperature. It was common knowledge that the total energy radiated by a body increased as the temperature increased, but this was made quantitative in 1879 by the Austrian physicist Josef Stefan (1835–1893). He showed that the total energy radiated by a body increased as the fourth power of the absolute temperature. (The absolute temperature, symbolized as °K, is equal to the centigrade temperature, °C, plus 273°; see page I–193.)

Consider a body, for instance, that is maintained at room temperature, 300°K, and is then radiating a certain amount of energy. If the temperature is raised to 600°K, which is that of melting lead, the absolute temperature has been doubled and the total amount of energy radiated is increased by 2^4, or 16 times. If the same body is raised to a temperature of 6000°K, which is that of the surface of the sun, it is at an absolute temperature twenty times as high as it was at room temperature, and it radiates 20^4, or 160,000, times as much energy.

In 1884, Boltzmann (who helped work out the kinetic theory of gases) gave this finding a firm mathematical foundation and showed that it applied, strictly, to black bodies only and that non-black bodies always radiate less heat than *Stefan's law* would require. Because of his contribution, the relationship is sometimes called the *Stefan-Boltzmann law*.

But it is not only the total quantity of energy that alters with rising temperature. The nature of the light waves emitted also changes, as is, in fact, the common experience of mankind. For objects at the temperature of a steam radiator, for instance (less than 400°K), the radiation emitted is in the low frequency infrared. Your skin absorbs the infrared and you feel the radiation as heat, but you see nothing. A radiator in a dark room is invisible.

As the temperature of an object goes up, it not only radiates more heat, but the frequency of the radiation changes somewhat, too. By the time a temperature of 950°K is reached, enough radiation, of a frequency high enough to affect the retina, is emitted for the body to appear a dull red in color. As the temperature goes higher still, the red brightens and eventually turns first orange and then yellow as more and more of still higher frequencies of light are emitted. At a temperature of 2000°K, an object, although glowing brightly, is still emitting radiation that is largely in the infrared. It is only when the temperature reaches 6000°K, the temperature of the surface of the sun, that the emitted radiation is chiefly in the visible light region of the spectrum. (Indeed, it is probably because the sun's surface is at that particular temperature, that our eyes have evolved in such a fashion as to be sensitive to that particular portion of the spectrum.)

Toward the end of the nineteenth century, physicists attempted to determine quantitatively the distribution of radiation among light of different frequencies at different temperatures. To do this accurately a black body was needed, for only then could one be sure that at each frequency all the light possible (for that temperature) was being radiated. For a non-black body, certain frequencies were very likely to be radiated in a deficient manner; the exact position of these frequencies being dependent on the chemical nature of the radiating body.

Since no actual body absorbs all the light falling upon it, no actual body is a true black body, and this seemed to interpose a serious obstacle in the path of this type of research. In the 1890's, however, a German physicist, Wilhelm Wien (1864–1928), thought of an ingenious way of circumventing this difficulty.

Imagine a furnace with a hole in it. Any light of any wavelength entering that hole would strike a rough inner wall and be mostly absorbed. What was not absorbed would be scattered in diffuse reflections that would strike other walls and be absorbed there. At each contact with a wall, additional absorption would take place, and only a vanishingly small fraction of the light would manage to survive long enough to be reflected out the hole again. That hole, therefore, would act as a perfect absorber (within reason) and would, therefore, represent a black body. If the furnace were raised to a certain temperature and maintained there, then the radiation emitted from that hole is *black-body radiation* and its frequency distribution can be studied.

In 1895, Wien made such studies and found that at a given temperature, the energy radiated at given frequencies, increased as the frequency was raised, reached a peak, and then began to decrease as the frequency was raised still further.

If Wien raised the temperature, he found that more energy was radiated at every frequency, and that a peak was reached again. The new peak, however, was at a higher frequency than the first one. In fact, as he continued to raise the temperature, the frequency peak of radiation moved continuously in the direction of higher and higher frequencies. The value of the peak frequency (ν_{max}) varied directly with the absolute temperature (T), so *Wien's law* can be expressed as follows:

$$\nu_{max} = kT \qquad \text{(Equation 8–1)}$$

where k is a proportionality constant.

Both Stefan's law and Wien's law are of importance in astronomy. From the nature of a star's spectrum, one can obtain a measure of its surface temperature. From this one can obtain a notion of the rate at which it is radiating energy and, therefore, of its lifetime. The hotter a star, the more short-lived it may be expected to be.

Wien's law explains the colors of the stars as a function of temperature (rather than as a matter of approach or recession as Doppler had thought—see page 76). Reddish stars are comparatively cool, with surface temperatures of 2000–3000°K. Orange stars have surface temperatures of 3000–5000°K, and yellow stars (like our sun) of 5000–8000°K. There are also white stars with surface temperatures of 8000–12,000°K and bluish stars that are hotter still.

Planck's Constant

At this point the paradox arose, for there remained a puzzle as to just why black-body radiation should be distributed in the manner observed by Wien. In the 1890's, physicists assumed that a radiating body could choose at random a frequency to radiate in. There are many more small gradations of high-frequency radiation than of low-frequency radiation (just as there are many more large integers than small ones), and if radiation could choose any frequency at random, many more high frequencies would be chosen than low ones.

Lord Rayleigh worked out an equation based on the assumption that all frequencies could be radiated with equal probability. He found that the amount of energy radiated over a particular range of frequencies should vary as the fourth power of the frequency. Sixteen times as much energy should be radiated in the form of violet light as in the form of red light, and far more still should be radiated in the ultraviolet. In fact, by Rayleigh's formula, virtually all the energy of a radiating body should be radiated very rapidly in the far ultraviolet. Some people referred to this as the "violet catastrophe."

The point about the violet catastrophe, however, was that it did not happen. To be sure, at very low frequencies the Rayleigh equation held, and the amount of radiation climbed rapidly as the frequency of the radiation increased. But soon the amount of radiation began to fall short of the prediction. It reached a peak at some intermediate frequency, a peak that was considerably below what the Rayleigh equation predicted for that frequency, and then, at higher frequencies still, the amount of radiation rapidly decreased, though the Rayleigh formula predicted a still-continuing increase.

· On the other hand, Wien worked up an equation designed to express what was actually observed at high frequencies. Unfortunately, it did not account for the distribution of radiation at low frequencies.

In 1899, a German physicist, Max Karl Ernst Ludwig Planck (1858–1947), began to consider the problem. Rayleigh's analysis, it seemed to Planck, was mathematically and logically correct, provided his assumptions were accepted; and since Rayleigh's equation did not fit the facts, it was necessary to question the assumptions. What if all frequencies were not, after all, radiated with equal probability? Since the equal-probability assumption required that more and more light of higher and higher frequency be radiated, whereas the reverse was observed, Planck proposed that the probability of radiation decreased as frequency increased.

Thus, there would be two effects governing the distribution of black-body radiation. First was the undeniable fact that there were more high frequencies than low frequencies so that there would be a tendency to radiate more high-frequency light than low-frequency light. Second, since the probability of radiation decreased as frequency went up, there would be a tendency to radiate less in the high-frequency range.

At very low frequencies, where the probability of radiation is quite high, the first effect is dominant and radiation increases as frequency rises, in accordance with the Rayleigh formula. However, as frequency continues to rise, the second effect becomes more and more important. The greater number of high frequencies is more than balanced by the lesser probability of radiating at such high frequency. The amount of radiation begins to climb more slowly as frequency continues to rise, reaches a peak, and then begins to decline.

Suppose the temperature is raised. This will not change the first effect, for the fact that there are more high frequencies than low frequencies is unalterable. However, what if a rise in temperature increased the probability that high-frequency light could be radiated? The second effect would therefore be weakened. In that case, radiation (at a higher temperature) could continue to increase, with higher frequencies, for a longer time before it was overtaken and repressed by the weakened second effect. The peak radiation consequently would move into higher and higher frequencies as the temperature went up. This was exactly what Wien had observed.

But how account for the fact that the probability of radiation decreased as frequency increased? Planck made the assumption that energy did not flow continuously (something physicists had always taken for granted) but was given off in discrete quantities. In other words, Planck imagined that there were "atoms of energy" and that a radiating body could give off one atom of energy or two atoms of energy, but never one and a half atoms of energy or, indeed, anything but an integral number of such entities. Furthermore, Planck went on to suppose, the energy content of such an atom of energy must vary directly with the frequency of the light in which it was radiated.

Planck called these atoms of energy, *quanta* (singular, *quantum*) from a Latin word meaning "how much?" since the size of the quanta was a crucial question.

Consider the consequences of this *quantum theory*. Violet light, with twice the frequency of red light, would have to radiate in quanta twice the size of those of red light. Nor could a quantum of violet light be radiated until enough energy had been accumulated to make up a full quantum, for less than a full quantum could not, by Planck's assumptions, be radiated. The probability, however, was that before the energy required to make up a full quantum of violet light was accumulated, some of it would have been bled off to form the half-sized quantum of red light. The higher the frequency of light, the less the probability that enough energy would accumulate to form a complete quantum without being bled off to form quanta of lesser energy content and lower frequency. This would explain why the "violet catastrophe" did not happen and why, in actual fact, light was radiated chiefly at low frequencies and more slowly than one might expect, too.

As the temperature rose, the general amount of energy available for radiation would increase as the fourth power of the absolute temperature. Under this increasing flood of radiation, it would become more and more likely that quanta of high-frequency light might have time to be formed. Thus, as Planck assumed, the probability of radiation in the high frequencies would increase, and the radiation peak would advance into higher frequencies. At temperatures of 6000°K, the peak would be in the visible light region, though the still larger quanta of ultraviolet would be formed in minor quantities even then.

If the energy content (e) of a quantum of radiation is proportional to the frequency of that radiation (v), we can say that:

$$e = hv \qquad \text{(Equation 8–2)}$$

where h is a proportionality constant, commonly called *Planck's constant*. If we solve Equation 8–2 for h, we find that $h = e/v$. Since the units of e in the cgs system are "ergs" and those of v are "1/seconds," the units of h are "ergs" divided by "1/seconds," or "erg-seconds." Energy multiplied by time is considered by physicists to be *action*. Planck's constant, therefore, may be said to be measured in units of action.

Planck derived an equation containing h that, he found, would describe the distribution of black-body radiation, as actually observed, over the entire range of frequencies. At least, it did this if h were given an appropriate, very small value. The best currently-accepted value of h is 0.00000000000000000000000000-00066256 erg-seconds or 6.6256 $\times 10^{-27}$ erg-seconds.

To see what this means, consider that orange light of wavelength 6000 A has a frequency of 50,000,000,000,000,000, or 5×10^{16} cycles per second. If this is multiplied by Planck's constant, we find that the energy content of a quantum of this orange light is $5 \times 10^{16} \times 6.6256 \times 10^{-27}$, or about 3.3×10^{-10} ergs. This is only about a third of a billionth of an erg, and an erg itself is but a small unit of energy.

It is no wonder, then, that individual quanta of radiant energy were not casually observed before the days of Planck.

Planck's quantum theory, announced in 1900, proved to be

a watershed in the history of physics. All physical theory that did not take quanta into account, but assumed energy to be continuous, is sometimes lumped together as *classical physics*, whereas physical theory that does take quanta into effect is *modern physics*, with 1900 the convenient dividing point.

Yet Planck's theory, when first announced, created little stir. Planck himself did nothing with it at first but explain the distribution of black-body radiation, and physicists were not ready to accept so radical a change of view of energy just to achieve that one victory. Planck himself was dubious and at times tried to draw his quantum theory as close as possible to classical notions by supposing that energy was in quanta form only when radiated and that it might be absorbed continuously.

And yet (with the wisdom of hindsight) we can see that quanta help explain a number of facts about absorption of light that classical physics could not. In Planck's time, it was well known that violet light was much more efficient than red light in bringing about chemical reactions, and that ultraviolet light was more efficient still. Photography was an excellent example of this, for photographic film of the type used in the nineteenth century was very sensitive to the violet end of the spectrum and rather insensitive to the red end. In fact, ultraviolet light had been discovered a century before Planck through its pronounced effect on silver nitrate (see page 77). Was it not reasonable to suppose that the large quanta of ultraviolet light could produce chemical reactions with greater ease than the small quanta of red light? And could not one say that this would only explain the facts if it were assumed that energy was absorbed only in whole quanta?

This argument was not used to establish the quantum theory in connection with absorption, however. Instead, Einstein made use of a very similar argument in connection with a much more recently-discovered and an even more dramatic phenomenon.

The Photoelectric Effect

In the last two decades of the nineteenth century, it had been discovered that some metals behave as though they were giving off electricity under the influence of light. At that time, physicists were beginning to understand that electricity was associated with the movement of subatomic particles called *electrons* and that the effect of light was to bring about the ejection of electrons from metal surfaces. This is the *photoelectric effect*.

On closer study, the photoelectric effect became a prime

puzzle. It seemed fair to assume that under ordinary conditions the electrons were bound to the structure of the metal and that a certain amount of energy was required to break this bond and set the electrons free. Furthermore, it seemed that as light was made more and more intense, more and more energy could be transferred to the metal surface. Not only would the electrons then be set free, but considerable kinetic energy would be available to them, so they would dart off at great velocities. The more intense the light, the greater the velocities. Nor did it seem that the frequency of the light ought to have anything to do with it; only the total energy carried by the light, whatever its intensity.

So it seemed, but that is not what happened.

The German physicist Philipp Lenard (1862–1947), after careful studies in 1902, found that for each surface that showed the photoelectric effect, there was a limiting *threshold frequency* above which, and only above which, the effect was to be observed.

Let us suppose, for instance, that this threshold frequency for a particular surface is 500 quadrillion cycles per second (the frequency of orange light of wavelength 6000 A). If light of lower frequency, such as red light of 420 quadrillion cycles per second, is allowed to fall upon the surface, nothing happens. No electrons are ejected. It doesn't matter how bright and intense the red light is and how much energy it carries; no electrons are ejected.

As soon, however, as the light frequency rises to 500 quadrillion cycles per second, electrons begin to be ejected, but with virtually no kinetic energy. It is as though the energy they have received from the light is just sufficient to break the bond holding them to the surface, but not sufficient to supply them with any kinetic energy in addition. Lenard found that increasing the intensity of the light at this threshold frequency did nothing to supply the electrons with additional kinetic energy. As a result of the increased intensity, more electrons were emitted from the surface, the number being in proportion to the energy of the orange light, but all of them lacked kinetic energy.

If the frequency were increased still further and if violet light of 1000 quadrillion cycles per second were used, electrons would be emitted with considerable kinetic energy. The number emitted would vary with the total energy of the light, but again all would have the same kinetic energy.

In other words, a feeble violet light would bring about the emission of a few high-energy electrons; an intense orange light would bring about the emission of many low-energy electrons; and an extremely intense red light would bring about the emission of no electrons at all.

The physical theories of the nineteenth century could not account for this, but in 1905 Einstein advanced an explanation that made use of Planck's quantum theory, which was now five years old but still very much neglected.

Einstein assumed that light was not only radiated in quanta, as Planck had maintained, but that it was absorbed in quanta also. When light fell upon a surface, the electrons bound to the surface absorbed the energy one quantum at a time. If the energy of that quantum was sufficient to overcome the forces holding it to the surface, it was set free—otherwise not.

If course, an electron might conceivably gain enough energy to break loose after absorbing a second quantum even if the first quantum had been insufficient. This, however, is an unlikely phenomenon. The chances are enormous that before it can absorb a second quantum, it will have radiated the first one away. Consequently, one quantum would have to do the job by itself; if not, merely multiplying the number of quantums (which remain individually insufficient) would not do the job. To use an analogy, if a man is not strong enough to lift a boulder single-handed, it doesn't matter if one million men, each as strong as the first, try one after the other to lift it single-handed. The boulder will not budge.

The size of the quantum, however, increases as frequency increases. At the threshold frequency, the quantum is just large enough to overcome the electron bond to a particular surface. As the frequency (and the energy content of the quantum) increase further, more and more energy will be left over, after breaking the electron bond, to be applied as kinetic energy.

For each substance, there will be a different and characteristic threshold energy depending on how strongly the electrons are bound to their substance. For a metal like cesium, where electrons are bound very weakly, the threshold frequency is in the infrared. Even the small quanta of infrared supply sufficient energy to break that weak bond. For a metal like silver, where electrons are held more strongly, the threshold frequency is in the ultraviolet.

Einstein suggested, then, the following relationship:

$$\tfrac{1}{2}mv^2 = h\nu - w \qquad\qquad \text{(Equation 8–2)}$$

where $\tfrac{1}{2}mv^2$ is the kinetic energy of the emitted electron; $h\nu$ (Planck's constant times frequency) the energy content of the quanta being absorbed by the surface; and w the energy required to break the electron free. At the threshold frequency, electrons would barely be released and would possess no kinetic energy. For that reason, Equation 8-2 would become $0 = hv - w$; and this would mean that $hv = w$. In other words, w would represent the energy of the light quanta at threshold frequency.

Einstein's explanation of the photoelectric effect was so elegant, and fit the observations so well, that the quantum theory sprang suddenly into prominence. It had been evolved, originally, to explain the facts of radiation, and now, without modification, it was suddenly found to explain the photoelectric effect, a completely different phenomenon. This was most impressive.

It became even more impresssive when in 1916 the American physicist Robert Andrews Millikan (1868–1953) carried out careful experiments in which he measured the energy of the electrons emitted by light of different frequency and found that the energies he measured fit Einstein's equation closely. Furthermore, by measuring the energy of the electrons ($\frac{1}{2}mv^2$), the frequency of the light he used (v), and the threshold frequency for the surface he was using (w), he was able to calculate the value of h (Planck's constant) from Equation 8-2. He obtained a value very close to that which Planck had obtained from his radiation equation.

Since 1916, then, the quantum theory has been universally accepted by physicists. It is now the general assumption that energy can be radiated or absorbed only in whole numbers of quanta and, indeed, that energy in all its forms is "quantized"—that is, can only be considered as behaving as though it were made up of indivisible quanta. This concept has offered the most useful views of atomic structure so far, as we shall see in Volume III of this book.

Photons

Einstein carried the notion of energy quanta to its logical conclusion. A quantum seemed to be analogous to an "atom of energy" or a "particle of energy," so why not consider such particles to be particles? Light, then, would consist of particles which were eventually called *photons* (from the Greek for "light").

This notion came as a shock to physicists. The wave theory of light had been established just a hundred years before and for a full century had been winning victory after victory, until Newton's particle theory had been ground into what had seemed complete oblivion. If light consisted of particles after all, what was to be done with all the evidence that pointed incontrovertibly to waves? What was to be done with interference experiments, polarization experiments, and so on?

The answer is that nothing has to be done to them. It is simply wrong to think that an object must be *either* a particle *or* a wave. You might just as well argue that *either* we are head-down and an Australian head-up, *or* we are head-up and an Australian head-down. A photon is *both* a particle *and* a wave, depending on the point of view. (Some physicists, half-jokingly, speak of "wavicles.") In fact, one can be more general than that (as I shall further explain in Volume III of this book) and insist that all the fundamental units of the universe are *both* particles *and* waves.

It is hard for a statement like that to sink in, for the almost inevitable response is: "But how can an object be both a particle and a wave at the same time?"

The trouble here is that we automatically try to think of unfamiliar objects in terms of familiar ones; we describe new phenomena by saying something such as "An atom is like a billiard ball" or "Light waves are like water waves." But this really means only that certain prominent properties of atoms or light waves resemble certain prominent properties of billiard balls or water waves. Not all properties correspond: an atom isn't as large as a billiard ball; a light wave isn't as wet as a water wave.

A billiard ball has both particle and wave properties. However, the particle properties are so prominent and the wave properties so obscure and undetectable that we think of a billiard ball as a particle only. The water wave is also both wave and particle, but here it is the wave properties that are prominent and the particle properties that are obscure. In fact, all ordinary objects are extremely unbalanced in that respect, so we have come to assume that an object must be either a particle or a wave.

The photons of which light is made up happen to be in better balance in this respect, with both wave properties and particle properties quite prominent. There is nothing in ordinary experience among particles and waves to which this can be compared. However, just because we happen to be at a loss for a familiar analogy, we need not think that a wave-particle is "against common sense" or "paradoxical" or, worse still, that "scientists cannot make up their minds."

We may see this more clearly if we consider an indirect analogy. Imagine a cone constructed of some rigid solid such as steel. If you hold such a cone point-upward, level with the eye, you will see its boundary to be a triangle. Holding it in that orientation (point-up), you will be able to pass it through a closely-fitting triangular opening in a sheet of steel, but not through a circular opening of the same area.

Next imagine the cone held point toward you and at eye-level. Now you see its boundary to be that of a circle. In that orientation it will pass through a closely-fitting circular opening in a sheet of steel, but not through a triangular opening of the same area.

If two observers, who were familiar with two-dimensional plane geometry but not with three-dimensional solid geometry, were conducting such experiments, one might hotly insist that the cone was triangular since it could pass through a triangular hole that just fit; the other might insist, just as hotly, that it was a circle, since it could pass through a circular hole that just fit. They might argue thus throughout all eternity and come to no conclusion.

If the two observers were told that both were partly wrong and both partly right and that the object in question had *both* triangular *and* circular properties, the first reaction (based on two-dimensional experience) might be an outraged, "How can an object be both a circle and a triangle?"

However, it is not that a cone *is* a circle and a triangle, but that it has both circular and triangular cross sections, which means that some of its properties are like those of circles and some are like those of triangles.

In the same way, photons are in some aspects wave-like and in others particle-like. The wave-like properties so beautifully demonstrated through the nineteenth century were the result of experiments that served to catch light in its wave-aspect (like orienting the cone properly in order to show it to be a triangle).

The particle-like properties were not so easily demonstrated. In 1901, to be sure, the Russian physicist Peter Nikolaevich Lebedev (1866–1911) demonstrated the fact that light could exert a very slight pressure. A mirror suspended in a vacuum by a thin fiber would react to this pressure by turning, and twisting the fiber. From the slight twist that resulted when a light beam shone on the mirror, the pressure could be measured.

Under some conditions, Lebedev pointed out, radiation pressure could become more important than gravitation. The frozen gases making up the surface of a comet evaporate as the comet approaches the sun, and the dust particles ordinarily held in place by the frozen gas are liberated. These particles are subjected to the comet's insignificant gravitational force and also to the pressure of the sun's tremendous radiation. The unusually large radiation pressure is greater than the unusually weak gravitation, and the dust particles are swept away (in part) by the radiation that is streaming outward in all directions from the sun.

It is this that causes a comet's tail, consisting as it does of light reflected from these dust particles, to be always on the side away from the sun. Thus, a comet receding from the sun is preceded by its tail. This orientation of the comet's tail caused the German astronomer Johannes Kepler (1571–1630) to speculate on the existence of radiation pressure three centuries before that existence could be demonstrated in the laboratory.

The existence of radiation pressure might ordinarily serve as an example of the particle properties of light, since we tend to think of such pressure as resulting from the bombardment of particles as in the case of gas pressure (see page I–199). However, in 1873, Maxwell (who had also worked on the kinetic theory of gases) had shown that there were good theoretical arguments in favor of the fact that light waves might, as waves and not as particles, exert radiation pressure.

A more clear-cut example of particle-like properties was advanced in 1922 by the American physicist Arthur Holly Compton (1892–1962). He found that in penetrating matter an X-ray (a very high-frequency form of light, to be discussed in some detail in Volume III of this book) sometimes struck electrons and not only exerted pressure in doing so, but was itself deflected! In being deflected, the frequency decreased slightly, which meant that the X ray had lost energy. The electron, on the other hand, recoiled in such a direction as to account for the deflection of the X ray, and gained energy just equal to that lost by the X ray. This deflection and energy-transfer was quite analogous to what would have happened if an electron had hit an electron or, for that matter, if a billiard ball had hit a billiard ball. It could not be readily explained by the wave theory. The *Compton effect* clearly demonstrated that an X ray photon could act as a particle. There were good reasons for supposing that the more energetic a photon, the more prominent its particle properties were compared to its wave properties. Therefore, the Compton effect was more easily demonstrated for an X ray photon than for the much less energetic photons of visible light, but the result was taken to hold for all photons. The particle-wave nature of photons has not been questioned since.

Whereas some experiments illuminate the wave properties of

light and some demonstrate its particle properties, no experiment has ever been designed which shows light behaving as both a wave and a particle simultaneously. (In the same way, a cone may be oriented so as to pass through a triangle, or so as to pass through a circle, but not in such a fashion as to pass through both.) The Danish physicist Niels Bohr (1865–1962) maintained that to design an experiment showing light to behave as both a wave and particle simultaneously not only *has not* been done but *cannot* be done in principle. This is called the *principle of complementarity*.

This is not really frustrating to scientists, though it sounds so. We are used to determining the overall shape of a solid object by studying it first from one side and then from another, and then combining, in imagination, the information so gathered. It does not usually occur to us to sigh at the fact that we cannot see an object from all sides simultaneously, or to imagine that only by such an all-sides-at-once view could we truly understand the object's shape. In fact, could we see all sides simultaneously, we might well be confused rather than enlightened, as when we first see a Picasso portrait intended to show a woman both full-face and profile at the same time.

If light is considered as having the properties of both a particle and a wave, there is certainly no need of a luminiferous ether, any more than we need an ether for gravitation or as a standard for absolute motion.

However much light may seem to be a wave form, in its transmission across a vacuum, it is the particle properties that are prominent. The photons stream across endless reaches of vacuum just as Newton had once envisaged his own less sophisticated particles to be doing.

Consequently, once relativity and quantum theory both came into general acceptance—say, by 1920—physicists ceased to be concerned with the ether.

Yet even if we consider light to consist of photons, it remains true that the photons have a wave aspect—that something is still waving. What is it that is waving, and is it anything material at all?

To answer that, let's take up the remaining two phenomena that, since ancient times, have been examples of what seemed to be action at a distance. It will take several chapters to do so, but the answer will eventually be reached.

CHAPTER 9

Magnetism

Magnetic Poles

Forces of attraction between bodies have undoubtedly been observed since prehistoric times, but (according to tradition, at least) the first of the ancient Greeks to study the attractive forces systematically was Thales (640?–546 B.C.).

One attractive force in particular seemed to involve iron and iron ore. Certain naturally occurring types of iron ore ("loadstone") were found to attract iron and, as nearly as the ancients could tell, nothing else. Thales lived in the town of Miletus (on the Aegean coast of what is now Turkey) and the sample of iron ore that he studied purportedly came from the neighborhood of the nearby town of Magnesia. Thales called it "ho magnetes lithos" ("the Magnesian rock") and such iron-attracting materials are now called *magnets*, in consequence, while the phenomenon itself is *magnetism*.

Thales discovered that amber (a fossilized resin called "elektron" by the Greeks), when rubbed, also exhibited an attractive force. This was different from the magnetic force, for whereas magnetism seemed limited to iron, rubbed amber would attract any light object: fluff, feather, bits of dried leaves. In later centuries, objects other than amber were found to display this property when rubbed, and in 1600 the English physician and physicist William Gilbert (1540–1603) suggested that all such objects be called "electrics" (from the Greek word for amber). From this, eventually, the word *electricity* came to be applied to the phenomenon.

Magnetism, while the more restricted force, seemed under the experimental conditions that prevailed in ancient and medieval times to be far the stronger. It was magnetism, therefore, that was the more thoroughly investigated in the two thousand years following Thales.

It was learned, for instance, that the property of magnetism could be transferred. If a sliver of steel is stroked with the naturally occurring magnetic iron ore, it becomes a magnet in its own right and can attract pieces of iron though previously it had not been able to do so.

Furthermore, if such a magnetized needle was placed on a piece of cork and set to floating on water, or if it was pivoted on an axis so that it might freely turn, it was discovered that the needle did not take any position at random, but oriented itself in a specific direction. That direction closely approximates the north-south line. Then, too, if one marks one end of the magnetized needle in some way, as by a notch or a small droplet of paint, it becomes quickly apparent that it is always the same end that points north, while the other end always points south.

Because the ends of the magnetized needle pointed, so it seemed, to the poles of the earth, it became customary to speak of the end that pointed north as the *north pole of the magnet,* and of the other as the *south pole of the magnet.*

It was bound to occur to men that if the north pole of a magnetized needle could really be relied upon to pivot in such a way as always to point north, an unexcelled method of finding direction was at hand. Until then, the position of the North Star by night and the position of the sun by day had been used, but neither would serve except in reasonably clear weather.

The Chinese were supposed to have made use of the magnetized needle as a direction finder in making their way across the trackless vastness of Central Asia. However, the first uses of the needle in ocean voyages are recorded among Europeans of the twelfth century. The needle was eventually mounted on a card on which the various directions were marked off about the rim of a circle. Since the directions encompassed the rim of the card, the magnetized needle came to be called a *compass.*

There is no doubt that the compass is one of those simple inventions that change the world. Men could cross wide tracts of ocean without a compass (some two thousand years ago the Polynesians colonized the islands that dotted the Pacific Ocean without the help of one), but a compass certainly helps. It is probably no accident that it was only after the invention of the compass that European seamen flung themselves boldly out into the Atlantic Ocean, and the "Age of Exploration" began.

The poles of a magnet are distinguished by being the points at which iron is attracted most strongly. If a magnetized needle is dipped into iron filings and then lifted free, the filings will cluster most thickly about the ends. In this sense, a magnet of whatever shape has poles that can be located in this manner. Nor do poles occur singly. Whenever a north pole can be located, a south pole can be located, too, and vice versa.

Nor is it difficult to tell which pole is the north and which the south, even without directly making a compass of the magnet. Suppose that two magnetized needles have been allowed to orient themselves north-south and that the north pole of each is identified. If the north pole of one magnet is brought near the south pole of the second magnet, the two poles will exhibit a mutual attraction, and if allowed to touch, will remain touching. It will take force to separate them.

On the other hand, if the north pole of one magnet is brought near the north pole of the other, there will be a mutual repulsion. The same is true if the south pole of one is brought near the south pole of the other. If the magnets are free to pivot about, they will veer away and spontaneously reorient themselves so that the north pole of one faces the south pole of the other. If north pole is forced against north pole or south pole against south pole, there will be a separation as soon as the magnets are released. It takes force to keep them in contact.

We might summarize this by saying: *Like poles repel; unlike poles attract.*

Once the north pole of a particular magnet has been identified, then, it can be used to identify the poles of any other magnet. Any pole to which it is attracted is a south pole. Any pole by which it is repelled is a north pole. This was first made clear as long ago as 1269 by one of the few experimentalists of the middle ages, the Frenchman Peter Peregrinus.

(In view of this, it might have been better if the north poles of magnets, attracted as they are in the direction of the North Pole, had been called south poles. However, it is too late to do anything about that now.)

It is easy to see that the force exerted by a magnetic pole varies inversely with difference. If one allows a north pole to approach a south pole, one can feel the force of attraction grow stronger. Similarly, if one pushes a north pole near another north pole, one can feel the force of repulsion grow stronger. The smaller the distance, the greater the force.

Of course, we cannot speak of a north pole by itself and a south pole by itself. Every north pole is accompanied by its south pole. Therefore, if a north pole of Magnet A is attracting the south pole of Magnet B, the south pole of Magnet A must be simultaneously repelling the south pole of Magnet B. This tends to complicate the situation.

If one uses long, thin magnets, however, this source of complication is minimized. The north pole of Magnet A is close to the south pole of Magnet B, while the south pole of Magnet A (at the other end of a long piece of metal) is considerably farther away. The south pole's confusing repulsive force is weakened because of this extra distance and may be the more safely ignored.

In 1785, the French physicist Charles Augustin de Coulomb (1736–1806) measured the force between magnetic poles at varying distances, using a delicate torsion balance for the purpose. Thus, if one magnetic needle is suspended by a thin fiber, the attraction (or repulsion) of another magnet upon one of the poles of the suspended needle will force that suspended needle to turn slightly. In doing so, it will twist the fiber by which it is suspended. The fiber resists further twisting by an amount that depends on how much it has already been twisted. A given force will always produce a given amount of twist, and from that amount of twist the size of an unknown force can be calculated. (Fifteen years later, Cavendish used such a balance to measure weak gravitational forces, see page I–50; a century later still, Lebedev used one to detect radiation pressure—see page 137.)

On making his measurements, Coulomb found that magnetic force varied inversely as the square of the distance, as in the case of gravitation. That is, the strength of the magnetic force fell to one-fourth its value when the distance was increased twofold, and the force increased to nine times its value when the distance was decreased to one-third its previous value. This held true whether the force was one of attraction or repulsion.

This can be expressed mathematically as follows: If the magnetic force between the poles is called F, the strength of the two poles, *m* and *m'*, and the distance between the two poles, *d*, then:

$$F = \frac{mm'}{d^2}$$ (Equation 9–1)

If the distance is measured in centimeters, then the force is determined in dynes (where 1 dyne is defined as 1 gram-centimeter per second per second, see page I–32). Suppose, then, that two poles of equal intensity are separated by a distance of 1 centimeter and that the force of magnetic attraction equals 1 dyne. It turns out then that $m = m'$, and therefore, $mm' = m^2$. Then, since both *F* and *d* have been set equal to 1, it follows from Equation 9–1 that under those conditions $m^2 = 1$ and, therefore, $m = 1$.

Consequently, one speaks of *unit poles* as representing poles of such strength that on being separated by 1 centimeter they exert a magnetic force (of either attraction or repulsion) of 1 dyne. In Equation 9–1, where *F* is measured in dynes and *d* in centimeters, *m* and *m'* are measured in unit poles.

If a magnetic pole of 5 unit poles exerts a force of 10 dynes on a unit pole at a certain point, the intensity of the magnetic force is 2 dynes per unit pole. One dyne per unit pole is defined as 1 *oersted* (in honor of the Danish physicist Hans Christian Oersted, whose contribution to the study of magnetism will be discussed on page 201). The oersted is a measure of magnetic force per unit pole or *magnetic intensity*, which is usually symbolized as *H*. We can say, then, that $H = F/m$, or:

$$F = mH$$ (Equation 9–2)

where F is magnetic force measured in dynes, *m* is magnetic strength in unit poles, and *H* is magnetic intensity in oersteds.

Magnetic Domains

In the existence of both north and south poles, and in the consequent existence of magnetic repulsion as well as magnetic attraction, there is a key difference between magnetism and gravitation. The gravitational force is entirely one of attraction, and no corresponding force of gravitational repulsion has yet been discovered.

For that reason, gravitational force is always at its ideal maximum without the existence of any neutralizing effects. A body possessing the mass of the earth will exert a fixed gravitational attraction whatever its temperature or chemical constitution.

On the other hand, magnetic attraction can always be neutralized to a greater or lesser extent by magnetic repulsion, so magnetic effects will occur only in certain kinds of matter and then in widely varying strengths.

One might suppose (and, as we shall see in Volume III of this book, the supposition is correct) that magnetism is widespread in nature and that magnetic forces exist in all matter. Matter might then be considered to consist of submicroscopic magnets. A point in favor of this view (at least in the case of iron and steel) is the fact, discovered early, that if a long magnetic needle is broken in two, both halves are magnets. The broken end opposite the original north pole becomes a south pole; the broken end opposite the original south pole becomes a north pole. This is repeated when each half is broken again and again. It is easy to imagine the original needle broken into submicroscopic lengths, each of which is a tiny magnet, each with a north pole and south pole.

These submicroscopic magnets, in most substances and under most conditions, would be oriented randomly, so there is little or no concentration of north poles (or south poles) in any one direction and therefore little or no detectable overall magnetic force. In some naturally occurring substances, however, there would be a tendency for the submicroscopic magnets to line up, at least to a certain extent, along the north-south line. There would then be a concentration of north poles in one direction and of south poles in the other; enough of a concentration to give rise to detectable magnetic force.

If, let us say, the north pole of such a magnet is brought near iron, the submicroscopic magnets in the iron are oriented in such a way that the south poles face the magnet and the north poles face away. The iron and the magnet then attract each other. If it is the south pole of the magnet that is brought near the iron, then the submicroscopic magnets in the iron are oriented in the opposite fashion, and again there is attraction. Either pole of a magnet will, for that reason, attract iron. While the iron is near the magnet or in contact with it so that its own magnetic components are oriented, it is itself a magnet. The process whereby iron is made a magnet by the nearness of another magnet is *magnetic induction*. Thus, a paper clip suspended from a magnet will itself attract a second paper clip which will attract a third, and so on. If the magnet is removed, all the paper clips fall apart.

Ordinarily, the submicroscopic magnets in iron are oriented with comparative ease under the influence of a magnet and are disoriented with equal ease when the magnet is removed. Iron usually forms a *temporary magnet*. If a sliver of steel is subjected to the action of a magnet, however, the submicroscopic magnets within the steel are oriented only with considerably greater difficulty. Once the magnet is removed from the steel, however, the disorientation is equally difficult—difficult enough not to take place under ordinary conditions, in fact; therefore, steel generally remains a *permanent magnet*.

Nor is it iron only that is composed of submicroscopic magnets, for it is not only iron that is attracted to a magnet. Other metals, such as cobalt and nickel (which are chemically related to iron) and gadolinium (which is not) are attracted by a magnet. So are a number of metal alloys, some of which contain iron and some of which do not. Thus, Alnico, which as the name implies is made up of aluminum, nickel and cobalt (plus a little copper), can be used to make more powerful magnets than those of steel. On the other hand, stainless steel, which is nearly three-fourths iron, is not affected by a magnet.

Nor need the magnetic substance be a metal. Loadstone itself is a variety of iron oxide, an earthy rather than a metallic substance. Since World War II, a whole new class of magnetic substances have been studied. These are the *ferrites*, which are mixed oxides of iron and of other metals such as cobalt or manganese.

A material that displays, or that can be made to display, a strong magnetic force of the sort we are accustomed to in an ordinary magnet is said to be *ferromagnetic*. (This is from the Latin "ferrum" meaning "iron," since iron is the best-known example of such a substance.) Nickel, cobalt, Alnico, and, of course, iron and steel are examples of ferromagnetic substances.

The question arises, though, why some materials are ferromagnetic and some are not. If magnetic forces are a common property of all matter (as they are), why cannot the submicroscopic magnets of pure copper or pure aluminum, for instance, be stroked into alignment by an already-existing magnet? Apparently, this alignment cannot be imposed from outside unless the substance itself cooperates, so to speak.

In ferromagnetic substances (but only under appropriate conditions even in those) there is already a great deal of alignment existing in a state of nature. The submicroscopic magnets tend to orient themselves in parallel fashion by the billions of billions, producing concentrations of north and south poles here and there within the iron. The regions over which magnetic forces are thus concentrated are called *magnetic domains*.

Iron and other ferromagnetic substances are made up of such magnetic domains, each of which is actually on the border of visibility. A finely divided powder of magnetic iron oxide spread over iron will tend to collect on the boundaries between adjacent domains and make them visible to the eye.

Despite the existence of these domains, iron as ordinarily produced is not magnetic. That is because the domains themselves are oriented in random fashion so that the magnetic force of one is neutralized by those of its neighbors. Therefore, stroking with an ordinary magnet does not orient the submicroscopic magnets themselves (this is beyond its power); it merely orients the domains. Thus, the ferromagnetic material has itself done almost all the work of alignment to begin with, and man proceeds to add one final touch of alignment, insignificant in comparison to that which is already done, in order to produce a magnet.

If a ferromagnetic substance is ground into particles smaller than the individual domains making it up, each particle will tend to consist of a single domain, or of part of one. The submicroscopic magnets within each will be completely aligned. If such a powder is suspended in liquid plastic, the domains can be aligned by the influence of a magnet easily and thoroughly by bodily rotation of the particles against the small resistance of the liquid (rather than against the much greater resistance of the iron itself in the solid state). By allowing the plastic to solidify while the system is still under the influence of the magnet, the domains will be permanently aligned, and particularly strong magnets will have been formed. Furthermore, such magnets can be prepared in any shape and can be easily machined into other shapes.

Anything which tends to disrupt the alignment of the domains will weaken or destroy the magnetic force of even a "permanent" magnet. Two magnets laid parallel, north pole to north pole and south pole to south pole, will, through magnetic repulsion, slowly turn the domains away, ruining the alignment and weaking the magnetic force. (That is why magnets should always be stacked north-to-south.) From a more mechanical standpoint, if a magnet is beaten with a hammer, the vibration will disrupt alignment and weaken the magnetic force.

In particular, increased atomic vibration, caused by a rise in temperature (see page I–203), will disrupt the domains. In fact, for every ferromagnetic substance there is a characteristic temperature above which the alignment of the domains is completely disrupted and above which, therefore, the substance will show no ferromagnetic properties.

This was first demonstrated by the French physicist Pierre Curie (1859–1906) in 1895, and is therefore called the *Curie point*. The Curie point is usually below the melting point of a substance, so liquids are generally not ferromagnetic. The Curie point for iron, for instance, is 760°C whereas its melting point is 1539°C. For cobalt the Curie point is considerably higher, 1130°C, while for gadolinium it is considerably lower, 16°C. Gadolinium is only ferromagnetic at temperatures below room temperature. The Curie point may be located at very low temperatures indeed. For the metal dysprosium, its value is about −188°C (85°K), so it is only at liquid air temperatures that dysprosium forms domains and becomes ferromagnetic.

In some substances, the submicroscopic magnets spontaneously align themselves—but not with north poles pointing all in the same direction. Rather, the magnets are aligned in parallel fashion but with north poles pointing in one direction in half the cases and in the other direction in the remainder. Such substances are *antiferromagnetic,* and because the magnetic forces of one type of alignment are neutralized by those of the other, the overall magnetic force is zero. It may be, however, that the structure of the substance is such that the magnets with north poles pointing in one direction are distinctly stronger than those with north poles pointing in the other. In this case, there is a considerable residual magnetic force, and such substances are called *ferrimagnetic.* (Note the difference in the vowel.)

The ferrites are examples of ferrimagnetic materials. Naturally, a ferrimagnetic material cannot be as strongly magnetic as a ferromagnetic material would be, since in the latter all the domains are, ideally, pointing in the same direction, while in the former a certain amount of neutralization takes place. Thus, ferrites display only about a third the maximum strength that a steel magnet would display.

The Earth as a Magnet

The manner in which a compass needle pointed north-and-south was a tantalizing fact to early physicists. Some speculated that a huge iron mountain existed in the far north and that the magnetized needle was attracted to it. In 1600, the English physicist William Gilbert (1540–1603) reported systematic experiments that led to a more tenable solution.

A compass needle, as ordinarily pivoted, can rotate only about a vertical axis and is constrained to remain perfectly horizontal. What if it is pivoted about a horizontal axis and can, if conditions permit, point upward or downward? A needle so pivoted (in the northern hemisphere) does indeed dip its north pole several degrees below the horizon and toward the ground. This is called *magnetic dip*.

Gilbert shaped a loadstone into a sphere and used that to stand for the earth. He located its poles and decided its south pole, which attracted the north pole of a compass needle, would be equivalent to the earth's Arctic region, while the other would be equivalent to the Antarctic.

The north pole of a compass needle placed in the vicinity of this spherical loadstone pointed "north" as might be expected. In the loadstone's "northern hemisphere," however, the north pole of a compass needle, properly pivoted, also showed magnetic dip, turning toward the body of the loadstone. Above the south pole in the loadstone's "Arctic region," the north pole of the compass needle pointed directly downward. In the loadstone's "southern hemisphere," the north pole of the compass needle angled up away from the body of the loadstone and above its "Antarctic region" pointed straight upward.

Gilbert felt that the behavior of the compass needle with respect to the earth (both in its north-south orientation and in its magnetic dip) was strictly analogous to its behavior with respect to the loadstone. He drew the conclusion that the earth was itself a spherical magnet with its poles in the Arctic and Antarctic. The compass needle located the north by the same force that attracted it to the pole of any other magnet. (It is this natural magnetism of the earth that slowly orients the domains in certain types of iron oxide and creates the magnetized loadstone—from which all magnetic studies previous to the nineteenth century have stemmed.)

It might easily be assumed that the earth's magnetic poles are located at its geographic poles, but this is not so. If they were, the compass needle would point to more or less true north, and it does not. In Gilbert's time (1580), for instance, the compass needle in London pointed 11° east of north. The angle by which the needle deviates from true north is the *magnetic declination*. It varies from place to place on the earth, and in any given place varies from year to year. The magnetic declination in London is now 8° west of north, and since Gilbert's time, the declination has been as great as 25° west of north. In moving from an eastward declination in the sixteenth century to a westward one now, there had to be a time when declination was, temporarily, zero, and when the compass needle, in London, pointed due north. This was/ true in 1657.

The variation in declination with change in position was first noted by Christopher Columbus (1451–1506) in his voyage of discovery in 1492. The compass needle, which had pointed distinctly east of north in Spain, shifted westward as he crossed the Atlantic Ocean, pointed due north when he reached midocean, and distinctly west of north thereafter. He kept this secret from his crew, for they needed only this clear example of what seemed the subversion of natural law to panic and mutiny.

The existence of magnetic declination and its variation from spot to spot on the earth's surface would be explained if the magnetic poles were located some distance from the geographic poles. This is actually so. The south pole of the earth-magnet (which attracts the north pole of the compass needle) is located in the far north and is called, because of its position, the *north magnetic pole*. It is now located just off Canada's Arctic shore, some 1200 miles from the geographic north pole. The *south magnetic pole* (the north pole of the earth-magnet) is located on the shores of Antarctica, west of Ross Sea, some 1200 miles from the geographic south pole.

The two magnetic poles are not quite on opposite sides of the earth, so the line connecting them (the *magnetic axis*) is not only at an angle of about 18° to the geographic axis, but also does not quite pass through the earth's center.

The fact that magnetic declination alters with time seems to

Earth's magnetic poles

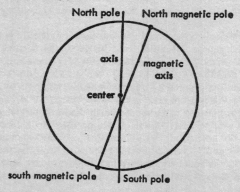

indicate that the magnetic poles change their position, and indeed the position of the north magnetic pole has shifted by several degrees from the time, one century ago, when it was first located.

For all its size, the earth is but a weak magnet. Thus, in even a small horseshoe magnet, the magnetic intensity between the poles can be as high as 1000 oersteds, yet the intensity of the earth's magnetism is only about 3/4 oersted even near the magnetic poles, where the intensity is highest. It sinks to 1/4 oersted at points equidistant from the magnetic poles (the *magnetic equator*).

Lines can be drawn on the face of the earth through points showing equal declination. These are called *isogonic lines* (from Greek words meaning "equal angles"). Ideally, they might be considered lines of "magnetic longitude." However, unlike geographic longitude, they are by no means arcs of great circles but curve irregularly in accordance with local magnetic properties of the earth's structure. And, of course, they change with time and must be constantly redrawn.

If it is agreed that the earth is a magnet, it yet remains to determine why it is a magnet. By the latter half of the nineteenth century, it came to seem more and more likely from several converging lines of evidence that the earth had a core of nickel-iron making up one third of its mass. Nothing seemed more simple than to suppose that this core was, for some reason, magnetized. However, it also seemed more and more likely that temperatures at the earth's core were high enough to keep that nickel-iron mass liquid and well above the Curie point. Therefore, the core cannot be an ordinary magnet, and the earth's magnetism must have a more subtle origin. This is a subject to which I will return.

Magnetic Fields

Magnetic force may fall off with distance according to an inverse square law (see Equation 9–1) as gravitational force does, but there are important differences. As far as we know, gravitational force between two bodies is not in the least affected by the nature of the medium lying between them. In other words, your weight is the same whether you stand on bare ground or insert an iron plate, a wooden plank, a foam rubber mattress or any other substance between yourself and the bare ground. For that matter, the sun's attraction of the earth is not altered when the 2000-mile thickness of the moon slips between the two bodies.

The force between magnetic poles does, however, alter with the nature of the medium between them, and Equation 9–1 holds strictly only where there is a vacuum between the poles. To explain this, we must consider the researches of the English scientist Michael Faraday (1791–1867).

In 1831, he noticed something that had been noticed over five centuries earlier by Peter Peregrinus and, undoubtedly, by numerous other men who had toyed with magnets through the centuries. . . . Begin with a sheet of paper placed over a bar magnet. If iron filings are sprinkled on the paper and the paper is then jarred so that the filings move about and take up some preferred orientation, those filings seem to follow lines that curve from one pole of the magnet to the other, crowding together near the poles and spreading apart at greater distances. Each line begins at one pole and ends at the other, and no two lines cross. (Of course, some of the lines seem incomplete because they run off the paper or because at great distances from the poles they are too weak to be followed accurately by the iron filings. Nevertheless it is reasonable to suppose that all the lines, however far they may have to sweep out into space and however weak they get, are nevertheless continuous from pole to pole.)

The shape of the lines depends on the shape of the magnet

Magnetic lines of force

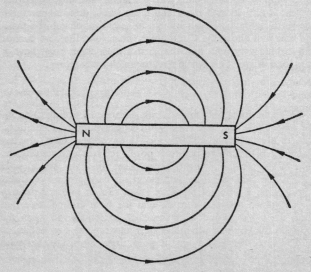

and the mutual relationship of the poles. In a horseshoe magnet, the lines cluster about the two poles and in the space between them are straight. The same is true if a north pole of one bar magnet is brought near the south pole of another. On the other hand, if the north pole of one bar magnet is brought near the north pole of another bar magnet, the lines of force curve away, those from one pole seeming to avoid touching those of the other.

Faraday called these *magnetic lines of force* and believed them to have real existence. He felt they were made of some elastic material that was stretched when extended between two unlike poles and that exerted a force tending to shorten itself, as an extended rubber band would do. It was this shortening tendency that accounted for magnetic attraction, according to Faraday.

The lines of force about a magnet of any shape, or about any system of magnets, can be mapped without the use of iron filings. A compass needle always orients itself in such a way as to lie along one of these lines. Therefore, by plotting the direction of the compass needle at various points in space, one can succeed in mapping the lines. In this way, the shape of the lines of force about the earth-magnet can be determined.

Faraday's view of the material existence of lines of force did not survive long. By mid-nineteenth century, the ether concept had grown strong in connection with light (see page 88ff.), and the magnetic lines of force came to be viewed as distortions of the ether.

With the disappearance of the ether concept at the start of the twentieth century, a further step had to be taken. Once again, it became a matter of the geometry of space itself. Suppose, for instance, you dropped a pencil into a cylindrical hollow. It would automatically orient itself parallel to the axis of the cylinder. If the cylinder were a tube that extended for miles, curving gently this way or that, a pencil dropped into it at any point would orient itself parallel to the axis of the tube, whatever the direction happened to be in that place. In fact, if you could not see the tube, but only the pencil, you could easily map the curves and sinuosities of the tube by noting the position taken up by the pencil at various points. The same is true of the compass needle and the magnetic lines of force.

Each magnetic pole affects the geometry of all of space, and this altered geometry (as compared with what the geometry would be in the absence of the magnetic pole) is called a *magnetic field*. The intensity of the magnetic field (the extent to which its geometry differs from ordinary nonmagnetic geometry of space) drops off as the square of the distance from the pole and soon becomes too small to detect. Nevertheless, the magnetic field of every magnetic pole in existence fills all of space, and the situation is made bearably simple only by the fact that in the immediate neighborhood of any given pole its field predominates over all others to such an extent that it may be considered in isolation. (This is true, of course, of gravitational fields as well.)

The concept of a magnetic field removes the necessity of supposing the magnetic force to represent action-at-a-distance. It is not that a magnet attracts iron over a distance but that a magnet gives rise to a field that influences a piece of iron within itself. The field (that is the space geometry it represents) touches both magnet and iron and no action-at-a-distance need be assumed.

Despite the fact that magnetic lines of force have no material existence, it is often convenient to picture them in literal fashion and to use them to explain the behavior of objects within a magnetic field. (In doing so, we are using a "model"—that is, a representation of the universe which is not real, but which aids thinking. Scientists use many models that are extremely helpful. The danger is that there is always the temptation to assume, carelessly, that the models are real, so they may be carried beyond their scope of validity. There may also arise an unconscious resistance to any changes required by increasing knowledge that cannot be made to fit the model.)

We can define the lines of force between two magnetic poles in the cgs system (using measurements in centimeters and dynes) in such a way that one line of force can be set equal to 1 *maxwell* (in honor of Maxwell, who did so much in connection with both gases and light). In the mks system, where the same measurements are made in meters and newtons, a line of force is set equal to 1 *weber* (in honor of the German physicist Wilhelm Eduard Weber [1804–1891]). The weber is much the larger unit, 1 weber being equal to 100,000,000 maxwells. Maxwells and webers are units of *magnetic flux*, a measurement you can imagine as representing the number of lines of force passing through any given area drawn perpendicularly to those lines.

What counts in measuring the strength of a magnetic field is the number of lines of force passing through an area of unit size. That is the *magnetic flux density*. The flux density measures how closely the lines of force are crowded together; the more closely they crowd, the higher the flux density and the stronger the magnetic field at that point. In the cgs system, the unit area is a square centimeter so that the unit of flux density is 1 maxwell per square centimeter. This is called 1 *gauss*, in honor of the German mathematician Karl Friedrich Gauss (1777–1855).* In the mks system, the unit area is a square meter and the unit of flux density is therefore 1 weber per square meter, a unit which has no special name. Since there are 10,000 square centimeters in a square meter and 100,000,000 maxwells in a weber, 1 weber per square meter is equal to 10,000 gausses.

Imagine a magnetic north pole and south pole separated by a vacuum. Lines of force run from pole to pole and the flux density at any given point between them would have a certain value, depending on the strength of the magnet. Is some material substance were placed between the poles then, even though the strength of the magnet were left unchanged, the flux density would change. The ratio of the flux density through the substance to that through a vacuum is called the *relative magnetic permeability*. Since this is a ratio, it is a pure number and has no units.

The permeability of a vacuum is set at 1 and for most material substances, the permeability is very nearly 1. Nevertheless, refined measurements show that it is never exactly 1, but is sometimes a little larger than 1 and sometimes a little smaller. Those substances with permeability a little larger than 1 are said to be *paramagnetic*, while those with permeability a little smaller than 1 are *diamagnetic*.

In a paramagnetic substance, with a permeability higher than 1, the flux density is higher than it would be in a vacuum. The lines of force crowd into the paramagnetic substance, so to speak, seeming to prefer it to the surrounding vacuum (or air). A paramagnetic substance therefore tends to orient itself with its longest axis parallel to the lines of force so that those lines of force can move through its preferred substance over a maximum distance. Again, since the flux density increases as one approaches a pole, there is a tendency for the paramagnetic substance to approach the pole (that is, to be attracted to it) in order that as many lines of force as possible can pass through it.

On the other hand, a diamagnetic substance, with a permeability of less than 1, has a flux density less than that of a vacuum (or air). The lines of force seem to avoid it and to crowd into the surrounding vacuum. Therefore a diamagnetic substance tends

* Most electrical and magnetic units are named in honor of scientists noted for their work in the field. Gauss and Weber established the first logical system of units in electricity and magnetism. Maxwell's contribution will be discussed on page 236 ff.

to orient itself with its longest axis perpendicular to the lines of force so that these lines of force need pass through the substance over only a minimum distance. Furthermore, the diamagnetic substance tends to move away from the pole (that is, be repelled by it) into a region of lower flux density so that as few lines as possible need pass through it.

Both effects are quite small and become noticeable only when very strong magnetic fields are used. The first to record such effects was Faraday, who found in 1845 that glass, sulfur and rubber were slightly repelled by magnetic poles and were therefore diamagnetic. The most diamagnetic substance known, at ordinary temperatures, is the element bismuth. (At extremely low temperatures, near that of absolute zero, the permeability of some substances drops to zero and diamagnetism is then at a maximum.)

Paramagnetism is considerably more common, and for a few substances permeability may be very high, even in the thousands. These high-permeability substances are those previously referred to as ferromagnetic. Here the attraction of the magnet and the orientation of iron filings parallel to the lines of force is so marked that it is easily noted.

Permeability (symbolized by the Greek letter "mu," μ) must be included in Coulomb's equation (Equation 9–2) to cover those cases where the poles are separated by more than vacuum:

$$F = \frac{mm'}{\mu d^2} \qquad\qquad \text{(Equation 9–3)}$$

Since μ is in the denominator, an inverse relationship is indicated. A diamagnetic substance with a permeability of less than 1 increases the magnetic force between poles, while a paramagnetic substance decreases the force. The latter effect is particularly marked when iron or steel, with permeabilities in the hundreds and even thousands, is between the poles. A bar of iron over both poles of a horseshoe magnet cuts down the magnetic force outside itself to such an extent that it almost acts as a magnetic insulator.

CHAPTER 10

Electrostatics

Electric Charge

Gilbert, who originated the earth-magnet idea, also studied the attractive forces produced by rubbing amber. He pivoted a light metal arrow so delicately that it would turn under the application of a tiny force. He could, in that way, detect very weak attractive forces and proceeded to find substances other than amber which, on rubbing, would produce such forces. Beginning in 1570, he found that a number of gems such as diamond, sapphire, amethyst, opal, carbuncle, jet, and even ordinary rock crystal produced such attractive forces when rubbed. He called these substances "electrics." A substance showing such an attractive force was said to be *electrified* or to have gained an *electric charge*.

A number of substances, on the other hand, including the metals in particular, could not be electrified and hence were "nonelectrics."

Eventually, electricity came to be considered a fluid. When a substance like amber was electrified, it was considered to have gained electric fluid which then remained there stationary. Such a charge was called *static electricity*, from a Latin word meaning "to be stationary," and the study of the properties of electricity under such conditions is called *electrostatics*.

Before electric forces could be studied easily, the fluid had to be concentrated in sizable quantities—in greater quantity than could be squeezed into small bits of precious and semiprecious materials. Some "electric" that was cheap and available in sizable quantities had to be found.

In the 1660's, the German physicist Otto von Guericke (1602–1686) found such a material in sulfur. He prepared a sphere of sulfur, larger than a man's head, arranged so that it could be turned by a crank. A hand placed on it as it turned gradually electrified it to a hitherto unprecedented extent. Guericke had constructed the first *electrical friction machine*.

Using it, Guericke discovered several similarities between electrostatic forces and magnetic forces. He found, for instance, that there was electrostatic repulsion as well as electrostatic attraction, just as there is both repulsion and attraction in the case of magnets. Again, a substance brought near the electrified sulfur itself exhibited temporary electrification, just as a piece of iron held near a magnet becomes itself temporarily magnetized. Thus there is *electrostatic induction* as well as magnetic induction.

In 1729, an English electrician, Stephen Gray (1696–1736), electrified long glass tubes and found that corks placed into the ends of the tubes, as well as ivory balls stuck into the corks by long sticks, became electrified when the glass itself was rubbed. The electric fluid, which came into being at the point of rubbing, must obviously spread throughout the substance, through the cork and the stick into the ivory, for instance. This was the first clear indication that electricity need not be entirely static but might move.

While the electric fluid, once it was formed within an "electric" by rubbing, might spread outward into every part of the substance, it would not pass bodily through it, entering at one point, for instance, and leaving at another. It was otherwise in the case of "nonelectrics," where such bodily passage did take place. Indeed, the flow of electric fluid took place extremely readily through substances like metals; so readily that a charged electric lost its charge altogether—was *discharged*—if it were brought into contact with metal that was in turn in contact with the ground. The fluid passed from the electric, via the metal, into the capacious body of the earth, where it spread out so thinly that it could no longer be detected.

That seemed to explain why metals could not be electrified by rubbing. The electric fluid, as quickly as it was formed, passed through the metal into almost anything else the metal touched. Gray placed metals on blocks of resin (which did not allow a ready passage to the electric fluid). Under such circumstances, pieces of metal, if carefully rubbed, were indeed electrified, for the fluid formed in the metal, unable to pass readily through the resin, was trapped, so to speak, in the metal. In short, as it eventually turned out, electric forces were universally present in matter, just as magnetic forces were.

As a result of Gray's work, matter came to be divided into two classes. One class, of which the metals—particularly gold, silver, copper and aluminum—were the best examples, allowed the passage of the electric fluid with great readiness. These are *electrical conductors*. The other group, such as amber, glass, sulfur, and rubber—just those materials that are easily electrified by rubbing—presents enormous resistance to the flow of electric fluid. These are *electrical insulators* (from a Latin word for "island," because such a substance can be used to wall off electrified objects, preventing the fluid from leaving and therefore making the objects an island of electricity, so to speak.)

Ideas concerning electrostatic attraction and repulsion were sharpened in 1733 by the French chemist Charles François Du Fay (1698–1739). He electrified small pieces of cork by touching them with an already electrified glass rod, so some of the electric fluid passed from the glass into the cork. Although the glass rod had attracted the cork while the latter was uncharged, rod and cork repelled each other once the cork was charged. Moreover, the two bits of cork, once both were charged from the glass, repelled each other.

The same thing happened if two pieces of cork were electrified by being touched with an already electrified rod of resin. However, a glass-electrified piece of cork attracted a resin-electrified piece of cork.

It seemed to Du Fay, then, that there were two types of electric fluid, and he called these "vitreous electricity" (from a Latin word for "glass") and "resinous electricity." Here, too, as in the case of the north and south poles of magnets, likes repelled and unlikes attracted.

This theory was opposed by Benjamin Franklin. In the 1740's, he conducted experiments that showed quite clearly that a charge of "vitreous electricity" could neutralize a charge of "resinous electricity," leaving no charge at all behind. The two types of electricity were therefore not merely different; they were opposites.

To explain this, Franklin suggested that there was only one electrical fluid, and all bodies possessed it in some normal amount.

When this fluid was present in its normal amount, the body was uncharged and showed no electrical effects. In some cases, as a result of rubbing, part of the electrical fluid was removed from the material being rubbed; in other cases it was added to the material. Where the body ended with an excess of the fluid, Franklin suggested, it might be considered *positively charged*, and where it ended with a deficit, it would be *negatively charged*. A positively-charged body would attract a negatively-charged body as the electric fluid strove (so to speak) to even out its distribution, and on contact, the electric fluid would flow from its place of excess to the place of deficiency. Both bodies would end with a normal concentration of the fluid, and both bodies would therefore be discharged.

On the other hand, two positively-charged bodies would repel each other, for the excess fluid in one body would have no tendency to add to the equal excess in the other—rather the reverse. Similarly, two negatively-charged bodies would repel each other.

Electrostatic induction was also easily explained in these terms. If a positively-charged object was brought near an uncharged body, the excess of fluid in the first would repel the fluid in the second and drive it toward the farthest portion of the uncharged body, leaving the nearest portion of that body negatively charged and the farthest portion positively charged. (The uncharged body would still remain uncharged, on the whole, for the negative charge on one portion would just balance the positive charge of the other.)

There would now be an attraction between the positively-charged body and the negatively-charged portion of the uncharged body. There would also be a repulsion between the positively-charged body and the positively-charged portion of the uncharged body. However, since the positively-charged portion of the uncharged body is farther from the positively-charged body than the negatively-charged portion is, the force of repulsion is weaker than the force of attraction, and there is a net attractive force.

The same thing happens if a negatively-charged body approaches an uncharged body. Here the electrical fluid in the uncharged body is drawn toward the negatively-charged body. The uncharged body has a positively-charged portion near the negatively-charged body (resulting in a strong attraction) and a negatively-charged portion farther from the negatively-charged body (resulting in a weaker repulsion). Again there is a net attractive force. In this way, it can be explained why electrically charged bodies of either variety attract uncharged bodies with equal facility.

Franklin visualized a positive charge and negative charge as being analogous to the magnetic north and south poles, as far as attraction and repulsion was concerned. There is an important difference, however. The magnetism of the earth offered a standard method of differentiating between the magnetic poles, depending upon whether a particular pole pointed north or south. No such easy way of differentiating a positive charge from a negative charge existed.

A positive charge, according to Franklin, resulted from an excess of electric fluid, but since there is no absolute difference in behavior between "vitreous electricity" and "resinous electricity," how could one tell which electric charge represents a fluid excess and which a fluid deficit? The two forms differ only with reference to each other.

Franklin was forced to guess, realizing full well that his chances of being right were only one in two—an even chance. He decided that glass, when rubbed, gained electric fluid and was positively charged; on the other hand, when resin was rubbed, it lost electric fluid and was negatively charged. Once this was decided upon, all electric charges could be determined to be either positive or negative, depending on whether they were attracted or repelled by a charge that was already determined to be either positive or negative.

Ever since Franklin's day, electricians have considered the flow of electric fluid to be from the point of greatest positive concentration to the point of greatest negative concentration, the process being pictured as analogous to water flowing downhill. The tendency is always to even out the unevenness of charge distribution, lowering regions of excess and raising regions of deficit.

Franklin's point of view implies that electric charge can neither be created nor destroyed. If a positive charge is produced by the influx of electric fluid, that fluid must have come from somewhere else, and a deficit must exist at the point from which it came. The deficit produced at its point of origin must exactly equal the excess produced at its point of final rest. Thus, if glass is rubbed with silk and if glass gains a positive charge, the silk gains an equal negative charge. The net electric charge in glass-plus-silk was zero before rubbing and zero after.

This view has been well substantiated since the days of Franklin, and we can speak of the *law of conservation of electric charge*.

We can say that net electric charge can neither be created nor destroyed; the total net electric charge of the universe is constant. We must remember that we are speaking of *net* electric charge. The neutralization of a quantity of positive electric charge by an equal quantity of negative electric charge is not the destruction of electric charge. The sum of $+x$ and $-x$ is 0, and in such neutralization it is not the net charge that has changed, only the distribution of charge. The same is true if an uncharged system is changed into one in which one part of the system contains a positive charge and another part an equal negative charge. This situation is exactly analogous to that involved in the law of conservation of momentum (see page I–67).

Electrons

Actually, both Du Fay's two-fluid theory of electricity and Franklin's one-fluid theory have turned out to possess elements of truth. Once the internal structure of the atom came to be understood, beginning in the 1890's (a subject that will be taken up in detail in Volume III of this book), it was found that subatomic particles existed, and that some of these possessed an electric charge while others did not.*

Of the subatomic particles that possess an electric charge, the most common are the *proton* and the *electron*, which possess charges of opposite nature. In a sense, then, the proton and the electron represent the two fluids of Du Fay. On the other hand, the proton, under the conditions of electrostatic experiments, is a completely immobile particle, while the electron, which is much the lighter of the two, is easily shifted from one body to another. In that sense, the electron represents the single electric fluid of Franklin.

In an uncharged body, the number of electrons is equal to the number of protons and there is no net charge. The body is filled with electric charge of both kinds, but the two balance. As a result of rubbing, electrons shift. One body gains an excess of electrons; the other is left with a deficit.

There is one sad point to be made, however. The electrons move in the direction opposed to that which Franklin had guessed for the electric fluid. Franklin had lost his even-money bet. Where

---

* Exactly what an electric charge *is*, we cannot say. However, we can say how a substance with an electric charge acts and how we may measure the extent of this action and, therefore, the size of the electric charge. This is an *operational definition* of electric charge, and it is enough to satisfy scientists, at least for the time being.

he thought an excess of electric fluid existed, there existed instead a deficit of electrons, and vice versa. For this reason, it was necessary to consider the electric charge of the electron to be negative; an excess of electrons would then produce the negative charge required by Franklin's deficiency of fluid, while a deficit of electrons would produce the positive charge of Franklin's excess of fluid. Since the electron is considered as having a negative charge, the proton must have a positive charge.

(Electrical engineers still consider the "electric fluid" to flow from positive to negative, although physicists recognize that electrons flow from negative to positive. For all practical purposes, it doesn't matter which direction of flow is chosen as long as the direction is kept the same at all times and there are no changes in convention in mid-stream.)

Coulomb, who measured the manner in which the force between magnetic poles was related to distance, did the same for the force between electrically charged bodies. Here his task was made somewhat easier because of an important difference between magnetism and electricity. Magnetic poles do not exist in isolation. Any body possessing a magnetic north pole must also possess a magnetic south pole. In measuring magnetic forces between poles, therefore, both attractive and repulsive forces exist, and they complicate the measurements. In the case of electricity, however, charges can be isolated. A body can carry a negative charge only or a positive charge only. For that reason, attractions can be measured without the accompaniment of complicating repulsions, and vice versa.

Coulomb found that the electric force, like the magnetic force, varied inversely as the square of the distance. In fact, the equation he used to express variation of electrical force with distance was quite analogous to the one he found for magnetic forces (see Equation 9-1, page 144).

If the electric charge on two bodies is q and q', and the distance between them is d, then F, the force between them (a force of attraction if the charges are opposite or of repulsion if they are the same) can be expressed:

$$F = \frac{qq'}{d^2}$$ (Equation 10-1)

provided the charges are separated by a vacuum.

In the cgs system, distances are measured in centimeters and forces in dynes. If we imagine, then, two equal charges separated by a distance of 1 centimeter and exerting a force of 1 dyne upon each other, each charge may be said to be 1 *electrostatic unit* (usually abbreviated *esu*) in magnitude. The esu is, therefore, the cgs unit of electric charge.

The smallest possible charge on any body is that upon an electron. Measurements have shown that to be equal to -4.8×10^{-10} esu, where the minus sign indicates a negative charge.* This means that a body carrying a negative charge of 1 esu contains an excess of approximately 2 billion electrons, while a body carrying a positive charge of 1 esu contains a deficit of approximately 2 billion electrons.

Another commonly used unit of charge—in the mks system —is the *coulomb*, named in honor of the physicist. A coulomb is equal to 3 billion esu. A body carrying a negative charge of 1 coulomb therefore contains an excess of approximately 6 billion billion electrons, while one carrying a positive charge of 1 coulomb contains a deficit of that many.

Imagine two electrons one centimeter apart. Since each has a charge of -4.8×10^{-10} esu, the total force (of repulsion, in this case) between them, using Equation 10–1, is $(-4.8 \times 10^{-10})^2$ or 2.25×10^{-19} dynes.

The two electrons also exert a gravitational force of attraction on each other. The mass of each electron is now known to be equal to 9.1×10^{-28} grams. The force of gravitational attraction between them is equal to Gmm'/d^2, where G is the gravitational constant which equals 6.67×10^{-8} dyne-cm^2/gm^2 (see page I–52). The gravitational force between the electrons is therefore equal to $(9.1 \times 10^{-28})^2$ multiplied by 6.67×10^{-8}, or 5.5×10^{-62} dynes.

We can now compare the strength of the electrical force and that of the gravitational force by dividing 2.25×10^{-19} by 5.5×10^{-62}. The quotient is 4×10^{42}, which means that the electrical force (or the comparable magnetic force in the case of magnets) is some four million trillion trillion trillion times as strong as the gravitational force. It is fair to say, in fact, that gravitational force is by far the weakest force known in nature.

The fact that gravitation is an overwhelming force on a cosmic scale is entirely due to the fact that we are then dealing with the huge masses of stars and planets. Even so, if we stop

---

* The proton has the same charge exactly, but a positive one. All subatomic particles have been found to have a charge exactly equal to that of an electron, or exactly equal to that of a proton, or to have no charge at all. Both negative and positive charges seem to come in packages of just one size and no other. Why that should be is a matter that is, as yet, unsolved.

to think that we, with our own puny muscles, can easily lift objects upward against the gravitational attraction of all the earth or, for that matter, that a small toy magnet can do the same, it must be borne in upon us that gravitational forces are unimaginably weak. And, in fact, when we deal with bodies of ordinary size, we completely neglect any gravitational forces between them.

Electrically charged objects serve as centers of *electric fields*, which are analogous to magnetic fields. There are *electric lines of force*, just as there are magnetic ones.

As in the case of magnetic lines of force, electric lines of force may pass through a material substance more readily, or less readily, than they would pass through an equivalent volume of a vacuum. The ratio of the flux density of electric lines of force through a medium to that through a vacuum is the *relative permittivity*. (This term is analogous to relative permeability in the case of magnetism.)

In general, insulators have a relative permittivity greater than 1; in some cases, much greater. The relative permittivity of air is 1.00054, while that of rubber is about 3, and that of mica about 7. For water, the value is 78. Where the relative permittivity is greater than 1, the electric lines of force crowd into the material and more pass through it than would pass through an equivalent volume of vacuum. For this reason, insulators are often spoken of as *dielectrics* (the prefix being from a Greek word meaning "through," since the lines of force pass through them). The relative permittivity is therefore frequently spoken of as the *dielectric constant*.

Coulomb's equation for the force between two charged particles might more generally be written, then:

$$F = \frac{qq'}{\kappa d^2} \qquad \text{(Equation 10–2)}$$

where the particles are separated by a medium with a dielectric constant of κ (the Greek letter "kappa").

Electric forces between charged particles decrease, then, if a dielectric is placed between; they decrease more as the dielectric constant is increased. The constituent particles of a substance like common table salt, for instance, are held together by electric attractions. In water, with its unusually high dielectric constant, these forces are correspondingly decreased, and this is one reason why salt dissolves readily in water (its particles fall apart, so to speak) and why water is, in general, such a good solvent.

Electromotive Force

If we rub a glass rod with a piece of silk, electrons shift from glass to silk; therefore, the glass becomes positively charged and the silk negatively charged. With each electron shift, the positive charge on the glass and the negative charge on the silk become higher, and it becomes increasingly difficult to move further electrons. To drag more negatively-charged electrons from the already positively-charged glass means pulling the electrons away against the attraction of the oppositely-charged glass. To add those electrons to the already negatively-charged silk means to push it on against the repulsion of like-charged bodies. As one proceeds to pile up positive charge on glass and negative charge on silk, the attraction and repulsion becomes larger and larger until, by mere hand-rubbing, no further transfer of electrons can be carried through.

This situation is quite analogous to that which arises in connection with gravitational forces when we are digging a hole. As one throws up earth to the rim of the hole, the level of earth around the rim rises while the level of earth within the hole sinks. The distance from the hole bottom to the rim top increases, and it becomes more and more of an effort to transfer additional earth from bottom to top. Eventually, the digger can no longer throw the shovelful of earth high enough to reach the height of the rim, and he has then dug the hole as far as he can.

This points up the value of using the familiar situations involving gravity as an analogy to the less familiar situations involving electric forces. Let us then, for a moment, continue to think of earth's gravitational field.

We can consider that a given body has a certain potential energy depending on its position with relation to earth's gravitational field (see page I–96). The higher a body (that is, the greater its distance from the earth's center), the greater its potential energy. To lift a body against earth's gravity, we must therefore add to its potential energy and withdraw that energy from someplace else (from our own muscles, perhaps). The quantity of energy that must be added, however, does not depend on the absolute value of the original potential energy of the body or of its final potential energy, but merely upon the difference in potential energy between the two states. We can call this difference in potential energy the *gravitational potential difference.*

Thus, an object on the 80th floor of a skyscraper has a greater potential energy than one on the 10th floor of the same skyscraper. All points on the 80th floor have the same potential energy, and all points on the 10th floor have the same potential energy. Both floors represent *equipotential surfaces*. To slide an object from one point on the 10th floor to another point on the 10th floor (ignoring friction) takes no energy since the gravitational potential difference is zero. The same is true in sliding an object from one point on the 80th floor to another point on the 80th floor. Though the absolute value of the potential energy on the 80th floor is greater, the gravitational potential difference is still zero.

Similarly, it is no harder to lift a body from the 80th to the 82nd floor than from the 10 to the 12th floor. (Actually, the gravitational force is a trifle weaker on the 80th floor than on the 10th, but the difference is so minute that it may be ignored.) It is the two-story difference that counts and that is the same in both cases. We can measure the difference in height (which is all that counts) by the amount of energy we must invest to raise a body of unit mass through that difference. In the mks system, the joule is the unit of energy (see page I–90) and the kilogram is the unit of mass. Therefore the unit of gravitational potential difference is a joule per kilogram.

There is an exact analogy between this and the situation in an electric field. Just as one adds energy to move one mass away from another mass, so must one add energy to move a negatively-charged body away from a positively-charged body, or vice versa. (One must also add energy to move a negatively-charged body toward another negatively-charged body or a positively-charged body toward another positively-charged body. For this there is no exact analogy in the gravitational system, since there is no such thing as gravitational repulsion.) The separation of unlike charged bodies or the approach of like charged bodies represents an increase in electric potential energy, and once the charged bodies have changed position with respect to each other, the difference in the electric potential energy is the *electric potential difference*. (The concept of a change in potential energy is so much more commonly used in electrical work than in other branches of physics that when the term *potential difference* is used without qualification, it may invariably be taken to refer to an electric potential difference rather than, say, to a gravitational one.)

Again the electric potential difference can be measured in terms of the energy that must be added to a unit charge to move it a given distance. In the mks system, the unit of charge is the coulomb so that the unit of electric potential difference is the joule per coulomb. This unit is used so often that a special name has been given to it, the *volt*, in honor of the Italian physicist Alessandro Volta (1745–1827), whose work will be described on page 178. As a result, the electric potential difference is sometimes referred to as the "voltage."

Let us return to the gravitational analogy again and consider an object resting on a flat surface. It has no tendency to move spontaneously to another portion of the flat surface, for there is a gravitational potential difference of zero between one point on the surface and another. On the other hand, if the object is suspended a meter above the surface and is released, it will spontaneously fall, moving from the point of higher potential energy to that of lower potential energy. It is the gravitational potential difference that brings about the spontaneous motion.

Similarly, an electric charge will have no spontaneous tendency to move from one point in an electric field to another point at the same potential energy level. If an electric potential difference exists, however, the electric charge will have a spontaneous tendency to move from the point of higher energy to that of lower. Since it is the electric potential difference that brings about spontaneous motion of electric charge, that potential difference can be spoken of as an *electromotive force* (a force that "moves electricity"), and this phrase is usually abbreviated as *emf*. Instead of speaking of a potential difference of so many volts, one frequently speaks of an emf of so many volts.

To create a potential difference, or an emf, in the first place, one must—in one way or another—bring about a separation of unlike charges or a crowding together of like charges. Thus, in rubbing a glass rod, one removes negatively-charged electrons from an increasingly positively-charged rod and adds negatively-charged electrons to an increasingly negatively-charged piece of silk.

It is sometimes possible to create an emf by squeezing certain crystals. A crystal is often made up of both positively- and negatively-charged particles arranged in orderly fashion in such a way that all the positively-charged particles and all the negatively-charged particles are grouped about the same central point. If two opposite faces of a crystal are placed under pressure, the crystal can be slightly flattened and distorted, and the charged particles making up the crystals are pushed together and spread out sideways. In most cases, both types of particles change position in identical fashion and remain distributed about the same central point. In some cases, however, the change is such that the average position of the negatively-charged particles shifts slightly with respect to the average position of the positively-charged particles. This means there is, in effect, a separation of positive and negative charges and a potential difference is therefore created between the two faces of the crystal.

This phenomenon was first discovered by Pierre Curie (who discovered the Curie point, see page 148) and his brother, Jacques, in 1880. They called the phenomenon, *piezoelectricity* ("electricity through pressure").

The situation can also be reversed. If a crystal capable of displaying piezoelectricity is placed within an electric field so that a potential difference exists across the crystal, the crystal alters its shape correspondingly. If the potential difference is applied and taken away, over and over again, the crystal can be made to vibrate and produce sound waves. If the crystal is of the proper size and shape, sound waves of such high frequency can be produced as to be in the ultrasonic range (see page I–180). Such interconversions of sound and electric potential are useful in today's record players.

Condensers

In working with electricity, it is sometimes convenient to try to place as much charge within a body as possible, with as little effort as possible. Suppose you have a metal plate, insulated in such a way that any electric charge added to it would remain. If you touch the plate with a negatively-charged rod, electrons will flow into the metal plate and give it a negative charge.

You can continue this process as long as you can maintain a potential difference between rod and plate—that is, as long as you can keep the rod, by protracted rubbing, more negatively charged than the plate. Eventually, however, you will increase the negative charge of the plate to such a level that no amount of rubbing will make the rod more negatively charged than that. The potential difference between rod and plate will then be zero, and a charge will no longer spontaneously move.

Suppose, however, you next bring a second metal plate, one that is positively charged, down over the first and parallel to it, but not touching. The electrons in the first plate are pulled toward the positively-charged second plate and crowd into the surface facing the positive plate. (The electrons crowding into that surface are now closer together than they were before, when they had been spread out evenly. They are "condensed" so to speak, and so this device of two flat plates held parallel and a short distance apart, may be called a *condenser*.)

With the electrons in the negative plate crowding into the surface facing the positive plate, the opposite surface has fewer electrons and a lower potential. There is once again a potential difference between the negatively-charged rod and that surface of the first plate which is away from the second plate. Electrons can once more pass from the rod into the plate, and the total charge on the plate can be built up considerably higher than would have been possible in the absence of the second plate.

Similarly, the positive charge on the second plate can be built up higher because of the presence of the negatively-charged first plate. Because the plates lend each other a greater capacity for charge, a condenser may also be called a *capacitor*.

The more highly charged the two plates (one positive and one negative), the greater the potential difference between them; this is the same as saying that the higher a mountain peak and the lower a valley, the greater the distance there is to fall. There is thus a direct relationship between the quantity of charge and the potential difference. If we imagine a vacuum between the plates, we can expect the ratio between charge and potential difference to be a constant, and we can express this as follows:

$$\frac{q}{v} = c \qquad \text{(Equation 10–3)}$$

where q is the charge in coulombs, and v is the potential difference in volts. The constant c is the *capacitance*, for which the units are coulombs per volt. One coulomb per volt is referred to as a *farad*, in honor of Michael Faraday.

Thus, a condenser (or capacitor) with a capacitance of 1 farad, will pile up a charge of 1 coulomb on either plate, one negative and one positive for every volt of potential difference between the plates. Actually, condensers with this large a capacitance are not generally met with. It is common, therefore, to use a *microfarad* (a millionth of a farad) or even a *micromicrofarad* (a millionth of a millionth of a farad) as units of capacitance.

Suppose, now, a dielectric (see page 165) is placed between the plates of a condenser. A dielectric decreases the force of attraction between given positive and negative charges (see page

165) and therefore lessens the amount of work required to separate these charges. But, as was explained on page 166, the potential difference is the measure of the work required to separate unlike charges. This means that the potential difference across the condenser, once the dielectric is placed between the plates, is v/κ, where κ is the dielectric constant.

If we call the capacitance of the condenser with the dielectric, c' then, in view of this:

$$c' = \frac{q}{v/\kappa} = \frac{\kappa q}{v} = \kappa \left(\frac{q}{v} \right) \qquad \text{(Equation 10—4)}$$

Combining Equations 10—3 and 10—4:

$$c' = \kappa c \qquad \text{(Equation 10—5)}$$

We see, then, that placing a dielectric between the plates of a condenser multiplies the capacitance of the condenser by the dielectric constant. The dielectric constant of air is only 1.0006 (where that of vacuum is taken as 1), so separation of the plates by air can be accepted as an approximation of separation by vacuum. The dielectric constant of glass is about 5, however, so if the plates of a condenser are separated by glass, its capacitance increases fivefold over the value for plates separated by air. For a given potential difference, a glass-separated condenser will pile up five times the charge an air-separated condenser will.

The capacitance can be further increased by reducing the distance between the plates or by increasing the area of the plates or both. If the distance between the plates is decreased, the potential difference decreases (as gravitational potential difference would decrease if two objects were one story apart instead of two stories apart). If this is so, then v in Equation 10—3 decreases while q remains unchanged and c necessarily increases. Again, if the plates were increased in area, there would be room for more electric charge to crowd in, so to speak. Consequently, q would increase in Equation 10—3 and so therefore would c.

A condenser with large plates can be unwieldy, but the same effect can be attained if one stacks a number of small condensers, and connects all the positive plates by a conducting material such as a metal rod, and all the negative plates by another metal rod. In that way, any charge added to one of the plates would spread out through all the plates of the same type, and the many small pairs of plates would act like one large pair. The condensers are said, in this case, to be connected *in series*.

In such a condenser, one set of plates can be fixed, while the other set is pivoted. By turning a knob connected to the rod about which the other set is pivoted, one can turn the negative plates, let us say, more and more into line with the positive. Essentially, only those portions of the plates which directly face each other have much condenser action. Consequently, as the pivoting set of plates moves more and more into line, the capacitance increases steadily. If the plates are turned out of line, the capacitance decreases. We have here a *variable condenser*.

An electrically charged object can be discharged if, for instance, a finger is placed to it and if the man attached to the finger is standing, without insulation, on the essentially uncharged ground—that is, if the man is *grounded*. If the object is negatively charged, electrons will flow from it through the man and into the earth until the negative charge is dissipated. If the object is positively charged, electrons will flow from the earth through the man and into the object until the positive charge is neutralized. In either case, there is a flow of charge through the body.

Since the sensations of a living body are mediated by the flow of tiny amounts of charge through the nerves, it is not surprising that the flow of charge that results from discharging a charged object can be sensed. If the flow of charge is a small one, the sensation may be no more than a tingle. If it is a large one, the sensation may be a strong pain like that produced by a sudden blow. One then speaks of an *electric shock*. (As in the case of a physical blow, a strong enough electric shock can kill.) Since condensers can pile up large charges of electricity, the shock received from such a condenser is much larger than that received by discharging an ordinary electrified rod of similar size.

This unpleasant property of condensers was discovered accidentally in 1745, when the first condensers were more or less casually brought into existence. This original condenser evolved into a glass jar, coated inside and outside with metal foil. It was corked and a metal rod pierced the cork. A metal chain, suspended from the rod, touched the metal foil inside the glass jar.

Suppose the metal foil outside the glass is grounded. If the metal rod sticking up from the cork is touched with a negatively-charged rod, electrons will enter the metal and spread downward into the internal foil coating. The negative charge on that internal foil repels electrons on the outer foil and forces them down the conductor connecting it to the ground, where those electrons spread into the general body of the planet and can be forgotten. If this is repeated over and over, a large negative charge is built up on the internal foil and a large positive charge on the external foil; a much larger charge (thanks to the fact that the layers of foil act as a glass-separated condenser) than the early experimenters could possibly have expected.

The first men to produce condensers of this sort (the German experimenter Ewald George von Kleist in 1745 and the Dutch physicist Pieter van Musschenbroek [1692–1761] in 1746) were surprised and even horrified when they discharged the devices and found themselves numbed and stunned by the shock. Von Kleist abandoned such experiments at once, and Van Musschenbroek proceeded further only with the greatest caution. Since Van Musschenbroek did his work at the University of Leyden, in the Netherlands, the condenser came to be called a *Leyden jar*.

Through the second half of the eighteenth century, the Leyden jar was used for important electrical experiments. A charge could be collected and then given off in such unprecedented quantities that it could be used to shock hundreds of people who were all holding hands, kill small animals, and so on. These experiments were not important in themselves but served to dramatize

Leyden jar

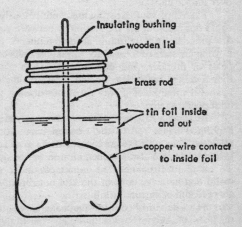

electrical phenomena and to rouse the interest of the scientific community (and of the general public, too).

In particular, the Leyden jar dramatized the matter of discharge through air. Dry air is an insulator, but insulation is never perfect, and if the charge on any object is great enough, it will force itself across a substance that is ordinarily an insulator. (Thus, you can imagine a weight resting upon a wooden plank suspended from its ends, and several feet above the ground. The wooden plank acts as an "insulator" in the sense that the weight cannot move downward despite its tendency to do so as a result of the gravitational potential difference between itself and the ground. If the weight is made heavier and heavier, a point will be reached where the plank breaks, and the weight drops. The "insulator" has been broken down, and the weight is "discharged," to use electrical terminology.)

When an electric charge forces itself across an ordinarily insulating gap of air, the air is heated by the electrical energy to the point where it glows. The discharge is therefore accompanied by a spark. The heated air expands and then, losing its heat to the surrounding atmosphere, contracts again. This sets up sound-wave vibrations, so the discharge is not only accompanied by a spark but also by a crackle. Such sparks and crackling were noted even by Guericke in his work with his charged ball of sulfur. With the Leyden jar and its much greater charge accumulation, sparks and crackling became most dramatic, and discharge would take place over longer gaps of air.

Franklin, who experimented industriously with Leyden jars, could not help but see the similarity between such a discharge and the thunder and lightning accompaniments to rainstorms. The Leyden jar seemed to produce miniature bolts of lightning and tiny peals of thunder, and contrariwise, earth and clouds during a thunderstorm seemed to be the plates of a gigantic Leyden jar. Franklin thought of a way of demonstrating that this was more than poetic fancy.

In June, 1752, he flew a kite during a thunderstorm. He tied a pointed metal rod to the wooden framework of the kite and attached a length of twine to it. This he attached to the cord that held the kite. He also attached an iron key to the end of the twine. To avoid electrocution, he remained under a shed during the storm and held the cord of the kite not directly, but by means of a dry length of insulating silk string.

The kite vanished into one of the clouds and Franklin noted the fibers of the kite cord standing apart as though all were charged and repelling each other. Presumably, the key had also gained a charge. Cautiously, Franklin brought the knuckle of his hand near the key; a spark leaped out, the same kind of spark, accompanied by the same crackle, one would expect of a Leyden jar. Franklin then brought out a Leyden jar he had with him and charged it with electricity from the clouds. The result was precisely the same as though he had charged it from an electrical friction machine. Franklin had thus, beyond doubt, showed that there was electricity in the high heavens just as there was on the ground, that lightning was a giant electrical discharge, and that thunder was the giant crackle that accompanied it.

He went further. Franklin experimented with the manner of discharge where bodies of different shapes were involved. Thus, if a metal sphere were brought near a charged body, there would be a discharge, let us say, across a one-inch gap of air. If a metal needle were brought near the same body charged to the same extent, discharge would take place across an airgap of six to eight inches. This could only be taken to mean that it was easier to discharge a charged body by means of a pointed object than a blunt object. Furthermore, the discharge by way of the point of a needle took place with such ease that it was not accompanied by noticeable sparks and crackles. (Nevertheless the fact of discharge could be detected easily enough, since the charged body suddenly lost the ability to repel a small similarly charged cork ball hanging in the vicinity.)

It occurred to Franklin that this phenomenon could be made use of on a large scale in connection with thunderstorms. If a long pointed metal rod were raised above the roof of a building, it would discharge the charge-laden thunderclouds more efficiently and quietly than would the building itself. The clouds would be discharged before they had built up enough charge to close the gap violently by means of a lightning bolt between themselves and the house. If conductors were attached to such a *lightning rod*, the charge drawn from the clouds could be conducted harmlessly to the earth and the house, in this manner, protected from lightning.

The lightning rod worked very well indeed, and over the next couple of decades, structures throughout America and Europe came under the protecting blanket of Franklin's invention. Franklin was the first great scientist produced by the New World and through this invention in particular he became famous among the scientists of Europe (a fact that had important political consequences when Franklin was sent on a mission to France during the American Revolution, a quarter-century after he flew his kite).

With the invention of the lightning rod, the study of electrostatics reached a climax. By the end of the eighteenth century, a new aspect of electricity came to the fore, and electrostatics receded into the background.

CHAPTER 11

Electric Currents

Continuous Electron Flow

Charge can move from one point to another (something which can also be described as the flowing of an electric current), as has been understood from the time of Gray in the early eighteenth century (see page 158). However, before 1800 only momentary flows of this sort were encountered. Charge could be transferred from a Leyden jar, for instance, to the human body, but after one quick spark, the transfer was done. A much huger charge transfer is that of lightning, yet "as quick as lightning" is a folk saying.

In order to arrange for a continuous transfer of charge, or a continuous flow of current from point A to point B, it is necessary to produce a new supply of charge at point A as fast as it is moved away, and consume it at point B as fast as it is brought there.

Methods for doing so developed out of the observations first made in 1791 by the Italian physician and physicist Luigi Galvani (1737–1798). Galvani was interested in muscle action and in electrical experiments as well. He kept a Leyden jar and found that sparks from it would cause the thigh muscles of dissected frogs to contract, even though there was no life in them. Others had observed this, but Galvani discovered something new: when a

177

metal scalpel touched the muscle at a time when a spark was drawn from a nearby Leyden jar, the muscle twitched even though the spark made no direct contact.

Suspecting this might be caused by induced electric charge in the scalpel, Galvani exposed frogs' thigh muscles to the electrified atmosphere of a thunderstorm, suspending them by brass hooks on an iron railing. He obtained his contractions but found that a thunderstorm was not, after all, necessary. All that was necessary was that the muscle be touched simultaneously by two different metals, whether any electric spark was in the vicinity or not, and whether there was a thunderstorm or not.*

Two dissimilar metals in simultaneous contact with a muscle could not only produce muscle contractions, but they could do so a number of times. It seemed certain that electricity was somehow involved and that whatever produced the electric charge was not put out of action after discharge and muscle contraction; instead the charge could be spontaneously regenerated again and again. Galvani made the assumption that the source of the electricity was in muscle and spoke of "animal electricity."

Others, nevertheless, suspected that the origin of the electric charge might lie in the junction of the two metals rather than in muscle, and outstanding among this group was the Italian physicist Alessandro Volta (1745–1827). In 1800, he studied combinations of dissimilar metals, connected not by muscle tissue but by simple solutions that by no stretch of the imagination could be considered to have any connection with a "life force."

He used chains of dissimilar metals, rightly suspecting that he could get better results from a number of sources combined than from a single one. He first used a series of bowls half full of salt water (each taking the place of a frog muscle) and connected them by bridges of metal strips, each composed of copper and zinc soldered together. The copper end was dipped into one bowl and the zinc end into another. Each bowl contained a copper end of one bridge on one side and the zinc end of another bridge on the other side.

Such a "crown of cups," as Volta called it, could be used as a source of electricity, which was thus clearly shown to originate in the metals and not in animal tissue. What's more, the electricity

---

* As a result of these experiments, which gained much public attention, a person reacting by muscle contractions to a shock of electricity (or to any unexpected sensation or even to a sudden strong emotion) is said to be "galvanized."

was produced continuously and could be drawn off as a continuous flow.

To avoid quantities of fluid that could slosh and spill, Volta tried another device. He prepared small discs of copper or silver (coins did very well) and other discs of zinc. He then piled them up: silver, zinc, silver, zinc, silver, zinc, and so on. Between each silver-zinc pair, he placed cardboard discs that had been moistened with salt water and that served the purpose of Galvani's frog muscle or Volta's own bowl of salt water. If the top of such a "voltaic pile" was touched with a metal wire, a spark could then be drawn out of the bottom, assuming that the bottom was touched with the other end of the same wire. In fact, if top and bottom were connected, a continuous current would flow through the wire.

The reason for this was not thoroughly understood for another century, but it rested on the fact that atoms of all matter include as part of their internal structure negatively-charged electrons and positively-charged protons. The electric charge produced by a continuously operating voltaic pile is therefore not truly created but is present constantly in matter. For a pile to work it is only necessary that it serve, in some manner, to separate the already-existing negative and positive charges.

Such separation is most simply described where two different metals alone are involved. Imagine two metals, say zinc and copper, in contact. Each metal contains electrons, bound by forces of greater or lesser extent to the atoms of the metal. The forces binding the electrons to zinc atoms are somewhat weaker than those binding electrons to copper. At the boundary, then, electrons tend to slip across from zinc to copper. The copper, with its stronger grip, wrests the electrons, so to speak, from the zinc.

This does not continue for long, for as the electrons enter the copper, that metal gains a negative charge while the zinc comes to have a positive charge. Further transfer of electrons away from the attraction of the positively-charged zinc and into the repelling force of the negatively-charged copper quickly becomes impossible, so an equilibrium is reached while the charge on each metal is still extremely minute. Still, the charge is large enough to be detected, and because unlike charges have been separated, a *contact potential difference* has been set up between the two metals.

If the temperature is changed, the force attracting electrons to atoms is also changed, but generally by different amounts for different metals. Imagine a long strip of zinc and a long strip of copper in contact at the two ends only (a *thermocouple*) and each end kept at a different temperature. There is a contact potential difference at each end, but the two have different values. The copper may be able to seize more electrons at end A than at end B, because at the temperature of end A, its electron-binding force has been strengthened to a greater extent than has that of the zinc.

Since the electron concentration in the copper at end A is greater than in the copper at end B, electrons flow through the copper from A to B. At end B there are now present too many electrons for the copper to retain at its particular temperature. Some of the electrons slip over to the zinc, therefore. Meanwhile, at end A, with some of the electrons lost, the copper can gain still more from the zinc.

The process continues indefinitely, with electrons traveling from end A to end B through the copper and then back from end B to end A through the zinc, and this continues as long as the temperature difference between the two ends is maintained. Such *thermoelectricity* was first observed in 1821 by the German physicist Thomas Johann Seebeck (1770–1831).

Possible practical applications of the phenomenon are not hard to see. The amount of current that flows through the thermocouple varies directly with the size of the temperature difference between the two ends; consequently, a thermocouple may be used as a thermometer. Indeed, if high melting metals such as platinum are used, thermocouples can be used to measure temperature in ranges far too high for ordinary thermometers. Furthermore, since even very minute electric currents can be easily detected and measured, thermocouples can be used to detect very feeble sources of heat; for example, that arising from the moon or from Venus.*

Chemical Cells

The junction of dissimilar metals by way of a conducting solution brings about a situation analogous to that of a thermocouple, but without the necessity of a temperature difference.

---

* Since an electric current can be maintained at the cost of a temperature difference alone, we can generate a continuous flow of electricity from burning kerosene (without moving parts). As we shall see in Chapter 12, other methods of forming electricity from burning fuel gained prominence in the course of the nineteenth century, but in the twentieth century, the potentialities of thermo-electricity are being reinvestigated. In particular, there is the possibility that if sunlight is used to maintain the temperature difference, solar energy may be converted to electricity and made use of directly, possibly in wholesale quantities. This would offer a most welcome addition to man's energy sources.

Suppose, for instance, that a strip of zinc is partially inserted into a solution of zinc sulfate. The zinc has a distinct tendency to go into solution. Each zinc atom, as it goes into solution, leaves two electrons behind so that the zinc rod gains a negative charge. The zinc atom minus two of the electrons it normally carries has a positive charge equal to that of the negative charge of the lost electrons. An electrically-charged atom is called an *ion*, so we may summarize matters by saying that the zinc in zinc sulfate produces positively-charged ions that enter the solution, while the zinc remaining behind gains a negative charge.

Imagine also a strip of copper inserted into a solution of copper sulfate. The copper sulfate solution contains positively-charged copper ions. There is no tendency for the copper metal to form more copper ions. Rather the reverse is true. The copper ions tend to return to the rod carrying with them their positive charge. Now suppose that the acid with its zinc strip and the copper sulfate with its copper strip are connected by a porous barrier so that liquid can slowly seep to and fro. We have a zinc strip carrying a small negative charge on one side and a copper strip carrying a small positive charge on the other.

If the two strips are connected by a wire, the surplus electrons in the negatively-charged zinc flow easily through the wire into the copper strip, which suffers a deficit of electrons. As the zinc loses its electron excess and therefore its negative charge, more zinc ions go into solution to produce a new electron excess. Moreover, as the copper gains electrons and loses its positive charge, more positively-charged copper ions can be attracted to the rod.

In short, electrons flow from the zinc to the copper by way of the wire, and then flow back from copper to zinc by way of the solution. The flow continues in its closed path until such time as all the zinc has dissolved or all the copper ions have settled out (or both). In the thermocouple, the electron flow was maintained by a temperature difference; in a voltaic pile, it was maintained by a chemical reaction.

Although the electron flow through the wire is from the zinc to the copper, electricians, following Franklin's wrong guess (see page 161), accept the convention that the flow of current is from the copper (the *positive pole*) to the zinc (the *negative pole*).

A generation after Volta's experiment, Faraday termed the metal rod that served as poles when placed in solutions, *electrodes*, from Greek words meaning "route of the electricity." The positive pole he called an *anode* ("upper route"), the negative pole a *cathode* ("lower route"), since he visualized the electricity as flowing downhill from anode to cathode.

Different chemicals so arranged as to give rise to a steady flow of electricity make up a *chemical cell*, or an *electrical cell*, or an *electrochemical cell*. All three names are used. Very often, as in Volta's original experiments, groups of cells are used. Groups of similar objects are referred to as "batteries," and for that reason groups of cells such as the voltaic pile are referred to as *electric batteries*, or simply *batteries*. (In ordinary conversation, even a single chemical cell may be referred to as a "battery.")

With Volta's discovery it became possible to study steady and long-continued flows of electric current. It was popular at first to call this phenomenon "galvanism" or "galvanic electricity," in honor of Galvani. However, it is more logical to call the study *electrodynamics* ("electricity-in-motion") as opposed to electrostatics. The study of those chemical reactions that give rise to electric currents is, of course, *electrochemistry*.

Electric currents were put to startling use almost at once. Since a flow of electrons is produced as a result of chemical reactions, it is not surprising that the electrons of a current, routed through a mixture of chemical substances, serve to initiate a chemical reaction. What's more, the chemical reactions that may easily be carried through by this method may be just those that prove very difficult to bring about in other ways.

In 1800, only six weeks after Volta's initial report, two English scientists, William Nicholson (1753–1815) and Anthony Carlisle (1768–1840), passed an electric current through water and found that they could break it up into hydrogen and oxygen. This process of bringing about a chemical reaction through an electric current is termed *electrolysis* ("loosening by electricity") because so often, as in the case of water, the reaction serves to break up a molecule into simpler substances.

In 1807 and 1808, the English chemist Humphry Davy (1778–1829), using a battery of unprecedented power, was able to decompose the liquid compounds of certain very active metals. He liberated the free metals themselves and was the first to form such metals as sodium, potassium, calcium, strontium, barium and magnesium—a feat that till then had been beyond the non-electrical abilities of chemists.

Davy's assistant, Faraday, went on to study electrolysis quantitatively and to show that the mass of substance separated by electrolysis was related to the quantity of electricity passing through the system. Faraday's *laws of electrolysis* (which will be taken up in some detail in Volume III) did much to help establish the atomistic view of matter then being introduced by the English chemist John Dalton (1766–1844). In the course of the next century, they helped guide physicists to the discovery of the electron and the working out of the internal structure of the atom.

As a result of Faraday's studies, a coulomb can be defined not merely in terms of total quantity of charge, or of total current (something not very easy to measure accurately), but as the quantity of current bringing about a certain fixed amount of chemical reaction (and that last could be measured quite easily). For instance, a coulomb of electric current passed through a solution of a silver compound will bring about the formation of 1.18 milligrams of metallic silver.

Chemists are particularly interested in a mass of 107.87 grams of silver for this is something they call "a gram-atomic weight of silver." Therefore, they are interested in the number of coulombs of current required to produce 107.87 grams of silver. But 107.87 grams is equal to 107,870 milligrams, and dividing that by 1.18 milligrams (the amount of silver produced by one coulomb), we find that it takes just about 96,500 coulombs to deposit a gram-atomic weight of silver out of a solution of silver compound. For this reason, 96,500 coulombs is referred to as 1 *faraday* of current.

A coulomb of electricity will deposit a fixed quantity of silver (or bring about a fixed amount of any particular chemical reaction) whether current passes through the solution rapidly or slowly. However, the rate at which the silver is deposited (or the reaction carried through) depends on the number of coulombs passing through the solution per unit of time. It would be natural to speak of the rate of flow of current (or of *current intensity*) as so many coulombs per second. One coulomb per second is called 1 *ampere*, in honor of the French physicist André Marie Ampère (1775–1836), whose work will be described on page 202. The ampere, then, is the unit of current intensity.

If, then, a current flowing through a solution of silver compound deposits 1.18 milligrams of metallic silver each second, we can say that 1 ampere of current is flowing through the solution.

Resistance

The rate of flow of electric current between point A and point B depends upon the difference in electric potential between these two points. If a potential difference of 20 volts serves to set up a current intensity of 1 ampere between those two points, a potential difference of 40 volts will produce a current intensity of 2 amperes, and a potential difference of 10 volts, one of 0.5 amperes.

This direct proportionality between potential difference and current intensity is true only if current is passed over a particular wire under particular conditions. If the nature of the path over which the current flows is changed, the relationship of potential difference and current intensity is changed, too.

Lengthening a wire, for instance, will reduce the current intensity produced in it by a given potential difference. If 20 volts will produce a current intensity of 1 ampere in a wire one meter long, the same 20 volts will produce only 0.5 amperes in a wire of the same composition and thickness but two meters long. On the other hand, if the wire is thickened, the current intensity produced by a given potential difference will be increased as the cross-sectional area or, which is the same thing, as the square of the diameter of the wire. If 20 volts will produce a 1-ampere current through a wire one millimeter in thickness, it will produce a 4-ampere current through an equal length of wire two millimeters in thickness.

Then, too, the nature of the substance conducting the electricity counts. If 20 volts produces a current of 3 amperes in a particular copper wire, it would produce a 2-ampere current in a gold wire of the same length and thickness, and a 1-ampere current through a tungsten wire of the same length and thickness. Through a quartz fiber of the same length and thickness, a current intensity of 0.00000000000000000000000003 amperes would be produced—so little that we might as well say none at all.

This sort of thing was investigated by the German physicist Georg Simon Ohm (1787–1854). In 1826, he suggested that the current intensity produced in a given pathway under the influence of a given potential difference depended on the *resistance* of that pathway. Doubling the length of a wire doubled its resistance; doubling its diameter reduced the resistance to a quarter of the original value. Substituting tungsten for copper increased resistance threefold, and so on.

The resistance could be measured as the ratio of the potential difference to the current intensity. If we symbolize potential difference as E (for "electromotive force"), current intensity as I, and resistance as R, then we can say:

$$R = \frac{E}{I}$$

(Equation 11–1)

This is *Ohm's law*. By transposition of terms, Ohm's law can of course also be written as $I = E/R$ and as $E = IR$.

Resistance is measured, as one might perhaps expect, in *ohms*. That is, if 1 volt of potential difference produces a current intensity of 1 ampere through some conducting pathway, then the resistance of that pathway is 1 ohm. From Equation 11–1, applied to the units of the terms involved, we see that 1 ohm can be defined as 1 volt per ampere.

Sometimes it is convenient to think of the *conductance* of a substance rather than of the resistance. The conductance is the reciprocal of the resistance, and the unit of conductance was (in a rare fit of scientific whimsy) defined as the *mho*, which is "ohm" spelled backward.

A pathway with a resistance of 1 ohm has a conductance of 1/1, or 1 mho. A resistance of 3 ohms implies a conductance of 1/3 mhos; a resistance of 100 ohms implies a conductance of 1/100 mhos, and so on. If we symbolize conductance by C, from Equation 11–1, we can say that:

$$C = \frac{1}{R} = \frac{I}{E}$$ (Equation 11–2)

so that 1 mho is equal to 1 ampere per volt.

For any given substance, resistance depends upon the length and diameter of the conducting pathway (among other things). In general, the resistance varies directly with length (L) and inversely with the cross-sectional area (A) of the pathway. Resistance is, therefore, proportional to L/A. If we introduce a proportionality constant, ρ (the Greek letter "rho"), we can say that:

$$R = \frac{\rho L}{A}$$ (Equation 11–3)

The proportionality constant ρ is the resistivity, and each substance has a resistivity characteristic for itself. If we solve for resistivity by rearranging Equation 11–3, we find that:

$$\rho = \frac{RA}{L}$$ (Equation 11–4)

Since in the mks system, the unit of R is the ohm, that of A is the square meter (or meter^2) and that of L is the meter. The unit of ρ, according to Equation 11–4, would be ohm-meter^2 per meter, or *ohm-meter*.

The better the conductor, the lower the resistivity. The best conductor known is the metal silver, which at 0°C has a resistivity of about 0.0000000152, or 1.52×10^{-8} ohm-meters. Copper is close behind with 0.0000000154, while gold and aluminum come next with 0.0000000227 and 0.0000000263 respectively. In general, metals have low resistivities and are therefore excellent conductors. Even Nichrome, an alloy of nickel, iron and chromium, which has an unusually high resistivity for a metal, has a resistivity of merely 0.000001 ohm-meters. This low resistivity of metals comes about because their atomic structure is such that each atom has one or two electrons that are loosely bound. Charge can therefore easily be transferred through the metal by means of these electrons.*

Substances with atomic structures such that all electrons are held tightly in place have very high resistivities. Even tremendous potential differences can force very little current through them. Substances with resistivities of over a million ohm-meters are therefore nonconductors. Maple wood has a resistivity of 300 million ohm-meters; glass a resistivity of about a trillion; sulfur one of a quadrillion; and quartz something like 500 quadrillion.

Between the low-resistivity conductors and the high-resistivity nonconductors, there is a group of substances characterized by moderate resistivities, higher than that of Nichrome but less than that of wood. The best-known examples are the elements germanium and silicon. The resistivity of germanium is 2 ohm-meters at 0°C and that of silicon is 30,000. Substances like germanium and silicon are therefore called *semiconductors*.

Notice that the resistivities given above are for 0°C. The value changes as temperature rises, increasing in the case of metals. One can picture matters this way. The electrons moving through a conductor are bound to find the atoms of the substance as barriers to motion, and some electrical energy is lost in overcoming this barrier. This energy loss is due to the resistance of the medium. If the temperature of the conductor rises, the atoms of the conductor vibrate more rapidly (see page I–203), and the electrons

---

* The movement of electrons is not quite the same as the flow of electricity. The electrons move at a certain not-very-high velocity, but the force that makes them move is transmitted much more quickly. If you flick a checker into a row of other checkers in contact, the incoming checker strikes and comes to a halt (perhaps rebounding a bit). The checkers it strikes remain in position for the most part, but the final checker at the opposite end of the line goes shooting off. The individual checkers scarcely moved, but the momentum was transmitted along the line of checkers at a speed that depends on the elasticity of the material making up those checkers. Similarly, quite apart from the actual velocity of electrons, the electric force travels through any substance at the speed of light.

find it more difficult to get through; consequently, resistivity rises. (Compare your own situation, for instance, making your way first through a quietly-standing crowd and then through the same crowd when it is milling about.)

If the resistivity is a given value at 0°C (ρ_o), it rises by a fixed small fraction of that amount ($\rho_o a$) for each degree rise in temperature (t). The increase in resistivity for any given temperature is therefore $\rho_o a t$. The total resistivity at that temperature (ρ_t) is therefore equal to the resistivity at 0°C plus the increase, or:

$$\rho_t = \rho_o + \rho_o a t = \rho_o (1 + at) \qquad \text{(Equation 11–5)}$$

The constant, a, which is the fractional increase in resistivity per degree, is the *temperature coefficient of resistance*.

As long as the temperature coefficient of resistance remains unchanged, the actual resistance of a particular conductor varies with temperature in a very simple way. From the resistance of a high-melting metal wire of given dimensions it is therefore possible to estimate high temperatures.

For semiconductors, the temperature coefficient of resistivity is negative—that is, resistance decreases as temperature goes up. The reason for this is that as temperature goes up the hold of the material on some of its electrons is weakened; more are available for moving and transferring charge. The increase in the number of available electrons more than makes up for the additional resistance offered by the more strongly vibrating atoms, so overall resistance decreases.

If the temperature coefficient of resistance of conductors was truly constant, one might expect that resistance would decrease to zero at temperatures just above absolute zero. However, at low temperatures, resistivity slowly decreases and the rate at which resistance declines as temperature drops slows in such a way that as the twentieth century opened, physicists were certain that a metal's resistance would decline to zero only at a temperature of absolute zero and not a whit before. This seemed to make sense since only at absolute zero would the metallic atoms lose their vibrational energy altogether and offer no resistance at all to the movement of electrons.

However, actual resistance measurements at temperatures close to absolute zero became possible only after the Dutch physicist Heike Kamerlingh-Onnes (1853–1926) managed to liquefy helium in 1908. Of all substances, helium has the lowest liquefaction point, 4.2°K, and it is only in liquid helium that the study of ultra-low temperatures is practical. In 1911, Kamerlingh-Onnes found to his surprise that the resistance of mercury, which was growing less and less in expected fashion as temperature dropped, suddenly declined precipitously to zero at a temperature of 4.16°K.

A number of other metals show this property of *superconductivity* at liquid helium temperatures. There are some alloys, in fact, that become superconductive at nearly liquid hydrogen temperatures. An alloy of niobium and tin remains superconductive up to a temperature of 18.1°K. Others, like titanium, become superconductive only at temperatures below 0.39°K. Although as many as 900 substances have been found to possess superconductive properties in the neighborhood of absolute zero, there remain many substances (including the really good conductors at ordinary temperatures, such as silver, copper, and gold) that have, as yet, shown no superconductivity at even the lowest temperatures tested.

Electric Power

It takes energy to keep an electric current in being against resistance. The amount of energy required varies directly with the total amount of current sent through the resistance. It also varies directly with the intensity of the current. Since, for a given resistance, the current intensity varies directly with potential difference (as is required by Ohm's law), we can say that the energy represented by a particular electric current is equal to the total quantity of charge transported multiplied by the potential difference.

Since energy can be transformed into work, let us symbolize the electric energy as W. This allows us to keep E for potential difference, and we can let Q stand for the total quantity of charge transported. We can then say:

$$W = EQ$$

(Equation 11–6)

The unit of potential difference is the volt and that of the total charge, the coulomb. If energy equals potential difference multiplied by total charge transferred, the unit of energy must equal volts multiplied by coulombs. However, a volt has been defined (see page 168) as a joule per coulomb. The units of energy must therefore be (joule/coulomb) (coulomb), or joules. The joule is the mks unit of energy, and we can say, then, that when 1 coulomb of electric charge is transported across a resistance under a potential difference of 1 volt, then 1 joule of energy is expended and may be converted into other forms of energy such as work, light, or heat.

It is often more useful to inquire into the rate at which energy is expended (or work performed) rather than into the total energy (or total work). If two systems both consume the same amount of energy or perform the same amount of work, but one does it in a minute and the other in an hour, the difference is clearly significant.

The rate at which energy is expended or work performed is termed *power*. If we consider the energy expended per second, the units of power become *joules per second*. One joule per second is defined as 1 *watt*, in honor of the Scottish engineer James Watt (1736–1819), whose work was dealt with on page I–93.

If 1 watt is equal to 1 joule per second and 1 joule is equal to 1 volt-coulomb (as Equation 11–6 indicates), then 1 watt may be considered as equal to 1 volt-coulomb per second. But 1 coulomb per second is equal to 1 ampere; therefore a volt-coulomb per second is equivalent to a volt-ampere, and we can conclude that 1 watt is equal to 1 volt-ampere.

What this means is that a current, driven by a potential difference of 1 volt and possessing an intensity of 1 ampere, possesses a power of 1 watt. In general, electric power is determined by multiplying the potential difference and the current intensity. If we symbolize the power as P, we can say:

$$P = EI \qquad \text{(Equation 11–7)}$$

An electrical appliance is usually rated in watts; what is indicated, in other words, is its rate of consuming electrical energy. We are most familiar with this in the case of light bulbs. Here the energy expended is used to increase the temperature of the filament within the bulb. The greater the rate of energy expenditure, the higher the temperature reached, and the more intense is the light radiated. It is for this reason that a 100-watt bulb is brighter than a 40-watt bulb—and hotter to the touch, too.

The potential difference of household current is usually in the neighborhood of 120 volts, and this remains constant. From Equation 11–7, we see that $I = P/E$. For a 100-watt bulb running on household current, then, $I = 100/120 = 5/6$. The current intensity in a 100-watt bulb is therefore 5/6 ampere. From this we can tell at once what the resistance (R) of the light bulb must be. Since, by Ohm's law, $R = E/I$, $R = 120$ divided by 5/6, or 144 ohms.

The watt is the unit of power in the mks system, but it is not the most familiar such unit. Quite frequently, one uses the *kilowatt*, which is equal to 1000 watts. Completely outside the mks system of units is the *horsepower*, which, in the United States at any rate, retains its popularity as the measure of the power of internal combustion engines. The horsepower is a larger unit than the watt; 1 horsepower = 746 watts. It follows that 1 kilowatt = 1.34 horsepower.

Since power is energy per time, energy must be power multiplied by time. This relationship (as always) carries over to the units. Since 1 watt = 1 joule/second, 1 joule = 1 watt-second. The *watt-second* is, therefore, a perfectly good mks unit of energy —as good as the joule to which it is equivalent. A larger unit of energy in this class is the *kilowatt-hour*. Since a kilowatt is equal to 1000 watts and an hour is equal to 3600 seconds, a kilowatt-hour is equal to (1000)(3600) watt-seconds, or joules. In other words, 1 kilowatt-hour = 3,600,000 joules. A 100-watt bulb (0.1 kilowatts) burning for 24 hours expends 2.4 kilowatt-hours of energy. The household electric bill is usually based on the number of kilowatt-hours of energy consumed.

From Ohm's law (Equation 11–1) we know that $E = IR$. Combining this with Equation 11–7, we find that:

$$P = I^2R$$

(Equation 11–8)

In other words, the rate at which energy is expended in maintaining an electric current varies directly with the resistance involved, and also with the square of the current intensity.

There are times when it is desirable to expend as little energy as possible in the mere transportation of current, as in conducting the current from the battery (or other point of origin) to the point where the electrical energy will be converted into some other useful form of energy (say the light bulb, where part of it will be converted into light). In that case, we want the resistance to be as small as possible. For a given length and thickness of wire, the lowest resistance is to be found in copper and silver. Since copper is much the cheaper of the two, it is copper that is commonly used as electrical wiring.

For really long distance transport of electricity, even copper becomes prohibitively expensive, and the third choice, the much cheaper aluminum, can be used. Actually, this is not bad at all, even though the resistivity of aluminum is 1.7 times as high as that of copper. The higher resistivity can be balanced by the fact that aluminum is only one-third as dense as copper, so a length of aluminum wire 1 millimeter in thickness is no heavier than the same length of copper wire 0.6 millimeters in thickness. Resistance decreases with increase in the cross-sectional area of the wire; consequently, the thicker aluminum wire actually possesses less resistance than does the same weight of thinner (and considerably more expensive) copper wire.

On the other hand, it is sometimes desired that as much of the electrical energy as possible be converted into heat, as in electric irons, toasters, stoves, driers, and so on. Here one would like to have the resistance comparatively high (but not so high that a reasonable current intensity cannot be maintained). Use is often made of high-resistance alloys such as Nichrome.

Within a light bulb, very high temperatures are particularly desired, temperatures high enough to bring about the radiation of considerable quantities of visible light (see page 126). There are few electrical conductors capable of withstanding the high temperatures required; one of these is tungsten. Tungsten has a melting point of 3370°C, which is ample for the purpose. However, its resistivity is only 1/20 that of Nichrome. To increase the resistance of the tungsten used, the filament in the light bulb must be both thin and long.

(At the high temperature of the incandescent tungsten filament, tungsten would combine at once with the oxygen of the air and be consumed. For this reason, light bulbs were evacuated in the early days of incandescent lighting. In the vacuum, however, the thin tungsten wires evaporated too rapidly and had a limited lifetime. To combat this, it became customary to fill the bulb with an inert gas, first nitrogen, and later on, argon. These gases did not react with even white-hot tungsten, and the gas pressure minimized evaporation and extended the life of the bulbs.)

Circuits

Suppose that a current is made to flow through a conductor with a resistance (R) of 100 ohms. Having passed through this, it is next led through one with a resistance (R') of 50 ohms, and then through one with a resistance (R'') of 30 ohms. We will speak of each of these items merely as "resistances," and for simplicity's sake we will suppose that the resistance of the conducting materials other than these three items is negligible and may be ignored.

Such resistances are *in series:* the entire current must pass first through one, then through the second, then through the third.

It is clear that the effect is the same as though the current had passed through a single resistance of $100 + 50 + 30$, or 180 ohms. Whenever items are connected in series so that a current passes through all, the total resistance is equal to the sum of the separate resistances.

If we are using household current with a potential difference of 120 volts, and if we assume that R, R' and R'' are the only resistances through which current is passing, then we can use Ohm's law to find the current intensity passing through the resistances. The total resistance is 180 ohms, and since $I = E/R$, the current intensity is equal to 120 volts divided by 180 ohms, or 2/3 amperes. All the current passes through all the resistances, and its intensity must be the same throughout.

Ohm's law can be applied to part of a system through which electricity is flowing, as well as to all of the system. For instance, what is the potential difference across the first of our three resistances, R? Its resistance is given as 100 ohms, and we have calculated that the current intensity within it (as well as within all other parts of the system in series) is 2/3 amperes. By Ohm's law, $E = IR$, so the potential drop across the first resistance is 100 ohms multiplied by 2/3 amperes, or 66 2/3 volts. Across the second resistance, R', it would be 50 ohms multiplied by 2/3 amperes, or 33 1/3 volts. Across the third resistance, R'', it is 30 ohms time 2/3 amperes, or 20 volts. The total potential difference is 66 2/3 + 33 1/3 + 20, or 120 volts.* Whenever items are in series, the total potential difference across all is equal to the sum of the potential differences across each separately.

Suppose that a fourth resistance is added to the series—one, let us say, of 60,000,000,000,000 ohms. The other resistances would add so little to this that they could be ignored. The current intensity would be 120 volts divided by 60 trillion ohms, or two trillionths of an ampere—an intensity so small we might just as well say that no current is flowing at all.

If two conductors are separated by a sizable air gap, current does not flow since the air gap has an extremely high resistance. For current to flow through conductors, there must be no significant air gaps. The current must travel along an unbroken path of reasonably conducting materials, all the way from one pole of a chemical cell (or other source of electricity) to the other pole.

---

* Even ordinary wiring, though it has as low a resistance as can be managed, does have some resistance. The wiring is in series with the objects to which it leads, and there is some potential difference across the wiring itself (*voltage drop*), though only to the extent of a volt or two, perhaps.

The pathway, having left the cell, must circle back to it, so one speaks of *electric circuits*.

If an air gap is inserted anywhere in a circuit made up of objects in series, a high resistance is added and the current virtually ceases. The circuit is "opened" or "broken." If the air gap is removed, the current flows again and the circuit is "closed."* Electric outlets in the wall involve no current flow if left unplugged, because an air gap exists between the two "terminals." This air gap is closed when an appliance is plugged in. An appliance is usually equipped, however, with an air gap within itself, so that even after it is plugged in, current will not flow. It is only when a switch is thrown or a knob is turned, and that second air-gap is also closed, that significant current finally starts to flow.

It may be desirable to form an air gap suddenly. There are conditions under which the current intensity through a particular circuit may rise to undesirable heights. As the current intensity goes up, the rate of energy expenditure and, therefore, the rate at which heat may develop, increases as the square of that intensity (see Equation 11–8). The heat may be sufficient to damage an electrical appliance or to set fire to inflammable material in the neighborhood.

To guard against this it is customary to include in the circuit, at some crucial point in series, a strip of low-melting alloy. A synonym for "melt" is "fuse," so such a low-melting material is a "fusible alloy." The little device containing a strip of such alloy is therefore an *electric fuse*. A rise in current intensity past some limit marked on the fuse (a limit of 15 amperes on one common type of household fuse, for instance) will produce enough heat to melt the alloy and introduce an air gap in the circuit. The current is stopped till the fuse is replaced. Of course, if the fuse is "blown" repeatedly, it is wise to have the circuit checked in order to see what is wrong.

When objects are in series within the circuit, the whole electric current passing through the first, passes through each of the succeeding objects consecutively. It is possible, though, that the current may have alternate routes in going from point A to point B, which may, for instance, be separately connected by the

* Naturally, if the potential difference is made high enough, current will flow with significant intensity through any resistance. The enormous potential differences between clouds, or between clouds and ground, during thunderstorms, is enough to transfer electric charge over tremendous air gaps, and man can duplicate this on a smaller scale ("man-made lightning"). However, ordinary circuits in everyday use run no risks of bursting across even small air gaps.

three different resistances, R, R', and R'', which I mentioned earlier as being of 100, 50 and 30 ohms respectively. The current flows in response to the potential difference between points A and B, and that has to be the same, regardless of the route taken by the current. (Thus, in going from the twelfth floor to the tenth floor of a building, the change in gravitational potential—two floors—is the same whether the descent is by way of the stairs, an elevator, or a rope suspended down the stairwell.)

Since under such circumstances the three resistances are usually shown, in diagrams, in parallel arrangement, they are said to be *in parallel*. We can then say that when objects are placed in parallel within an electric circuit, the potential drop is the same across each object.

It is easy to calculate the current intensity in each resistance under these circumstances, since the potential difference and resistance are known for each. If household current is used, with a potential difference of 120 volts, then that is the potential difference across each of the three resistances in parallel. Since by Ohm's law, $I = E/R$, the current intensity in the first resistance is 120/100, or 1.2 amperes, that in the second is 120/50, or 2.4 amperes, and that in the third is 120/30, or 4 amperes.

As you see, there is an inverse relationship between current intensity and resistance among objects in parallel. Since the reciprocal of resistance is conductance ($C = 1/R$), we can also say that the current intensity in objects in parallel is directly proportional to the conductances of the objects.

Imagine, a long wire stretching from point A to point B and arranged in a nearly closed loop so that points A and B, though separated by several meters of wire, are also separated by 1 centimeter of air gap. The wire and the air gap may be considered as being arranged in parallel. That is, current may flow from A to B through the long wire or across the short air gap. The resistance of the column of air between the points is, however, much greater than that of the long wire, and only a vanishingly small current intensity will be found in the air between the two points. For all practical purposes, all the current flows through the wire.

If the air gap is made smaller, however, the total resistance of the shortening column of air between A and B decreases, and more and more current intensity is to be found there. The passage of current serves to knock electrons out of atoms in the air, thus increasing the ability of air to conduct electricity by means of those electrons and the positively-charged ions the electrons leave behind. As a result, the resistance across the air gap further decreases. At some crucial point, this vicious cycle of current causing ions causing more current causing more ions builds rapidly to the point where current can be transferred through the air in large quantity, producing the spark and crackle that attracted such attention in the case of the Leyden jar (see page 173). Since the current takes the shortcut from point A to B, we speak of a *short circuit*. The wire from A to B, plus any appliances or other objects in series along that wire, is no longer on the route of the current, and the electricity has been shut off.

When a short circuit takes place so that a sizable portion of what had previously been the total circuit is cut out, there is a sudden decline in the total resistance of the circuit. The resistance across the sparking air gap is now very low, probably much lower than that of the wire and its appliances. There is a correspondingly higher current intensity in what remains of the circuit, and consequently, more heat develops. The best that can then happen is that the fuse blows. If there is any delay in this, then the sparks forming across the air gap may well set fire to anything inflammable that happens to be in the vicinity.

To minimize the possibility of short circuits, it is customary to wrap wires in insulation—silk, rubber, plastic, and so on. Not only do these substances have higher resistivities than air, but being solid, they set limits to the close approach of two wires, which remain always separated (even when pressed forcefully together) by the thickness of the insulation. When insulation is worn, however, and faults or cracks appear, short circuits become all too possible.

If we return to our three resistances in parallel, we might ask what the total resistance of the system is. We know the total current intensity in the system, since that is clearly the sum of the current intensities in each part of the system. The total current intensity in the particular case we have been considering would be $1.2 + 2.4 + 4.0 = 7.6$ amperes. The potential difference from A to B over any or all the routes in parallel is 120 volts. Since, by Ohm's law, $R = E/I$, the total resistance from A to B is 120 volts divided by the total current intensity, or 7.6 amperes. The total resistance, then, is 120/7.6, or just a little less than 16 ohms.

Notice that the total resistance is less than that of any one of the three resistances taken separately. To see why this should be, consider the Ohm's law relationship, $R_t = E/I_t$, as applied to a set of objects in parallel. Here R_t represents the total resistance and I_t the total current intensity; E, of course, is the same whether one of the objects or all of them are taken. The current intensity is equal to the sum of the current intensities (I, I' and I'') in the individual items. We can then say that:

$$R_t = \frac{E}{I + I' + I''}$$ (Equation 11–9)

If we take the reciprocal of each side, then:

$$\frac{1}{R_t} = \frac{I + I' + I''}{E} = \frac{I}{E} + \frac{I'}{E} + \frac{I''}{E}$$ (Equation 11–10)

By Ohm's law, we would expect I/E to equal R, I'/E to equal R', and I''/E to equal R'', the individual resistances of the items in parallel. Therefore:

$$\frac{1}{R_t} = \frac{1}{R} + \frac{1}{R'} + \frac{1}{R''}$$ (Equation 11–11)

We are dealing with reciprocals in Equation 11–11. We can say, for instance, that the reciprocal of the total resistance is the sum of the reciprocals of the individual resistances. It so happens that the lower the value of a quantity, the higher the value of its reciprocal, and vice versa. (Since 11 is greater than 3, 1/11 is smaller than 1/3, for example.) Thus, since the reciprocal of the total resistance ($1/R_t$) is the sum of the reciprocals of the separate resistances and therefore larger than any of the reciprocals of the separate resistances, the total resistance itself (R_t) must be smaller than any of the separate resistances themselves.

An important property of arrangements in parallel is this: If there is a break in the circuit somewhere in the parallel arrangement, the electricity ceases only in that branch of the arrangement in which the break occurs. Current continues to flow in the remaining routes from A to B. Because parallel arrangements are very common, a given electric outlet can be used even though all others remain open. Parallel arrangements also explain why a light bulb can burn out (forming an air gap in place of the filament) without causing all other light bulbs to go out.

Batteries

Throughout the first half of the nineteenth century, the chief source of electric current was the chemical cell, and though this has long since yielded pride of place as far as sheer work load is concerned, it remains popular and, indeed, virtually irreplaceable for many special tasks.

A very common type of electric cell used nowadays has as

its negative pole a cup of metallic zinc, and as its positive pole, a rod of carbon embedded in manganese dioxide.* Between the two is a solution of ammonium chloride and zinc chloride in water. Starch is added to the solution in quantities sufficient to form a stiff paste so that the solution will not flow and the cell is made "unspillable." The unspillability is so impressive a characteristic that the device is commonly called a *dry cell*. It is also called a "flashlight battery" because it is so commonly used in flashlights. It may even be called a *Leclanche cell* because the zinc-carbon combination was first devised in 1868 by the French chemist Georges Leclanché (1839–1882), though it was not until twenty years later that it was converted into its "dry" form.

The potential difference between the positive and negative poles of a chemical cell depends upon the nature of the chemical reactions taking place—that is, on the strength of the tendency of the substances making up the poles to gain or lose electrons. In the case of the dry cell, the potential difference is, ideally, 1.5 volts.

The potential difference can be raised if two or more cells are connected in series—that is, if the positive pole of one cell is connected to the negative pole of the next cell in line. In that case, current flowing out of the first cell under the driving force of a potential difference of 1.5 volts enters the second cell and gives that much of a "push" to the current being generated there. The current put out by the second cell is therefore driven by its own potential difference of 1.5 volts plus the potential difference of the cell to which it is connected—3.0 volts altogether. In general, when cells are connected in series so that all the current travels through each one, the total potential difference is equal to the sum of the potential differences of the individual cells.

Cells might also be connected in parallel—that is, all the positive poles would be wired together and all the negative poles would be wired together. The total current does not pass through each cell. Rather, each cell contributes its own share of the current and receives back its own share, so the potential difference of one has no effect on that of another. There are advantages, however, in having 1.5 volts supplied by ten cells rather than by one. For one thing, the total amount of zinc in ten cells is ten times greater than in one, and the ten-cell combination will continue delivering current for ten times as long.

Then, too, there is the question of *internal resistance* of a

* This is the case where the cell is cylindrical in shape. The two poles are arranged somewhat differently where the cell is box-shaped.

battery. When current is flowing, it flows not only through the wires and devices that make up the circuit connecting positive pole to negative pole, it must also flow from pole to pole within the cell by way of the chemical reactions proceeding there. The internal resistance is the resistance to this electric flow within the cell. The greater the current intensity withdrawn from the cell, the greater, too, the current intensity that must flow through the cell. The potential difference required to drive this current through the cell depends on the current intensity, for (by Ohm's law) $E = IR$. R, in this case, is the internal resistance of the cell, and E is the potential difference driving the current from negative to positive pole (using the electrician's convention). This potential difference is the direction opposite to that driving the current from the positive pole to the negative in the *external circuit* outside the cell, so the internal potential difference must be subtracted from the external one. In short, as you draw more and more current intensity from a cell, the potential difference it delivers drops and drops, thanks to internal resistance.

When cells are arranged in series, the internal resistance of the series is the sum of the internal resistance of the individual cells. The potential difference may go up, but ten cells in series will be as sensitive to high current intensities as one cell would be. When cells are arranged in parallel, however, the total internal resistance of the cells in the system is less than that of any single cell included, just as in the case of ordinary resistances (see page 196). A cell system in parallel can therefore deliver larger current intensities without significant drop in potential difference than a single cell could, although the maximum potential difference is no higher.

Electric cells of various sorts have been an incalculable boon to technological advance and are still extremely useful. Not only flashlights but a variety of devices from children's toys to radios can be powered by electric cells. Chemists like Davy even used them for important scientific advances requiring the delivery of fairly large amounts of electrical power. However, the really large-scale uses of electricity, such as that of powering huge factories and lighting whole cities, simply cannot be done by piling together millions of electric cells. The expense would be prohibitive.

The dry cell, for instance, obtains its energy by converting metallic zinc to zinc ions. Chemically, this is the equivalent of burning zinc—of using zinc as a fuel. When a dry cell is delivering 1 ampere of current, it consumes 1.2 grams of zinc in one hour. In that one hour, the power delivered by the battery would be 1.5 volts times 1 ampere, or 1.5 watts. Therefore 1.5 watt-hours is equivalent to the consumption of 1.2 grams of zinc, and 1 kilowatt-hour (1000 watt-hours) is equivalent to the consumption of 800 grams of zinc. A typical modern American household would, at the rate it consumes electricity, easily consume eight tons of zinc per year if dry cells were its source of supply (to say nothing of the other components involved.) Not only would this be ridiculously expensive, but the world's production of zinc could not maintain an economy in which individual families consumed the metal at this rate. In fact, it is fair to say that our modern electrified world simply could not exist on a foundation of ordinary chemical cells.*

One way of reducing the expense might be to devise some method of reversing the chemical reactions in a cell so that the original pole-substances might be used over again. This is not practical for the dry cell, but chargeable batteries do exist. The most common variety is one in which the negative pole is metallic lead and the positive pole is lead peroxide. These are separated by a fairly strong solution of sulfuric acid.

When such a cell is discharging, and an electric current is drawn off from it (at a potential difference of about 2 volts for an individual cell), the chemical reactions that proceed within it convert both the lead and the lead peroxide into lead sulfate. In the process, the sulfuric acid is consumed, too. If electricity is forced back into the cell (that is, if the negative pole of an electric source, working at a potential difference of more than 2 volts, is connected to the negative pole of the cell and the positive pole of the source is connected to the positive pole of the cell so that the cell is forced to "work backward" by a push that is stronger than its own), the chemical reaction goes into reverse. Lead and lead peroxide are formed once again, and the sulfuric acid solution grows stronger. The cell is "recharged." Such a cell was first devised in 1859 by the French physicist Gaston Planté (1834–1889).

In a superficial sense, it would seem that as the cell is re-

* Attempts are in progress to devise cells based on substances we would consider as more normal varieties of fuel. Such cells would be based on the combination of hydrogen, methane, alcohol, or even coal, with oxygen. In some cases bacterial action is used to bring about the necessary chemical reactions. Such *fuel cells* would be much cheaper than the more familiar chemical cells. By making it possible to draw electricity from burning fuel directly—and with greater efficiency than is now possible by methods to be described in the next two chapters—such cells could revolutionize various sectors of our economy. Fuel cells are only at the experimental stage, however, and were certainly not available during the decades, at the end of the nineteenth century, when the electrification of the industrialized regions of the earth began.

charged, electricity pours into the cell and is stored there. Actually, this is not so. Electricity is not directly stored; instead a chemical reaction is carried through, producing chemicals that can then react to generate electricity. Thus it is chemical energy that is stored, and such chargeable cells are called *storage batteries*. It is these (usually consisting of three to six lead-plus-lead-peroxide cells in series) that are present under the hood of automobiles.

Such a storage battery is heavy (because of the lead), dangerous to handle (because of the sulfuric acid), and expensive. Nevertheless, because it can be recharged over and over, the same battery can be used for years without replacement under conditions where heavy demands are periodically made upon it. Its usefulness, therefore, is not to be denied.

Yet where does the electricity come from that is used to recharge the storage battery? If such electricity is drawn from ordinary non-rechargeable cells, we are back where we started from. Clearly, in order to make it possible to use storage batteries on a large scale, the electricity used to charge it must have a much cheaper and more easily available source. In the automobile, for instance, the storage battery is continually being recharged at the expense of the energy of burning gasoline—which, while not exactly dirt-cheap, is certainly much cheaper and more available than the energy of burning zinc.

To explain how it is that burning gasoline can give rise to electric power, we must begin with a simple but crucial experiment conducted in 1819.

CHAPTER 12

Electromagnetism

Oersted's Experiment

Until the beginning of the nineteenth century, electricity and magnetism had seemed two entirely independent forces. To be sure, both were strong forces, both showed repulsion as well as attraction, and both weakened with distance according to an inverse square law. On the other hand, magnetism seemed to involve only iron plus (weakly) a few other substances, while electricity seemed universal in its effects; magnetism displayed poles in pairs only, while electricity displayed them in separation; and there was no flow of magnetic poles as there was a flow of electric charge. The balance seemed to be more in favor of the differences than the likenesses.

The matter was settled in 1819, however, as a result of a simple experiment conducted in the course of a lecture (and without any expectation of great events to come) by a Danish physicist, Hans Christian Oersted (1777–1851). He had been using a strong battery in his lecture, and he closed by placing a current-carrying wire over a compass in such a way as to have the wire parallel to the north-south alignment of the compass needle. (It is not certain now what point he was trying to make in doing this.)

However, when he put the wire over the needle, the needle turned violently, as though, thanks to the presence of the current, it now wanted to orient itself east-west. Oersted, surprised, carried the matter further by inverting the flow of current—that is, he connected the wire to the electrodes in reverse manner. Now the compass needle turned violently again, but in the opposite sense.

As soon as Oersted announced this, physicists all over Europe began to carry out further experimentation, and it quickly became plain that electricity and magnetism were intimately related, and that one might well speak of *electromagnetism* in referring to the manner in which one of the two forces gave rise to the other.

The French physicist Dominique François Jean Arago (1786–1853) showed almost at once that a wire carrying an electric current attracted not only magnetized needles but ordinary unmagnetized iron filings, just as a straightforward magnet would. A magnetic force, indistinguishable from that of ordinary magnets, originated in the electric current. Indeed, a flow of electric current *was* a magnet.

To show this more dramatically, it was possible to do away with iron, either magnetized or unmagnetized, altogether. If two magnets attracted each other or repelled each other (depending on how their poles were oriented), then the same should be true of two wires, each carrying an electric current.

This was indeed demonstrated in 1820 by the French physicist Ampère, after whom the unit of current intensity was named. Ampère began with two parallel wires, each connected to a separate battery. One wire was fixed, while the other was capable of sliding toward its neighbor or away from it. When the current was traveling in the same direction in both wires, the movable wire slid toward the other, indicating an attraction between the wires. If the current traveled in opposite directions in the two wires, the movable wire slid away, indicating repulsion between the two wires. Furthermore, if Ampère arranged matters so that the movable wire was free to rotate, it did so when the current was in opposite directions, turning through 180° until the two wires were parallel again with the current in each now flowing in the same direction. (This is analogous to the manner in which, if the north pole of one small magnet is brought near the north pole of another, the second magnet will flip so as to present its south pole end to the approaching north pole.)

Again, if a flowing current is a magnet, it should exhibit magnetic lines of force as an ordinary magnet does, and these lines of force should be followed by a compass needle. Since the compass needle tends to align itself in a direction perpendicular to that of the flow of current in the wire (whether the needle is held above or below the wire, or to either side), it would seem that the magnetic lines of force about a current-carrying wire appear in the form of concentric cylinders about the wire. If a cross section is taken perpendicularly through the wire, the lines of force will appear as concentric circles. This can be demonstrated by running a current-carrying wire upward through a small hole in a horizontal piece of cardboard. If iron filings are sprinkled on the cardboard and the cardboard is tapped, the filings will align themselves in a circular arrangement about the wire.

In the case of an ordinary magnet, the lines of force are considered to have a direction——one that travels from a north pole to a south pole. Since the north pole of a compass needle always points to the south pole of a magnet, it always points in the conventionally accepted direction of the lines of force. The direction of the north pole of a compass needle also indicates the direction of the lines of force in the neighborhood of a current-carrying wire, and this turns out to depend on the direction of the current-flow.

Ampère accepted Franklin's convention of current-flow from the positive electrode to the negative electrode. If, using this convention, a wire were held so that the current flowed directly toward you, the lines of force, as explored by a compass needle, would be moving around the wire in counterclockwise circles. If the current is flowing directly away from you, the lines of force would be moving around the wire in clockwise circles.

As an aid to memory, Ampère advanced what has ever since been called the "right-hand screw rule." Imagine yourself holding the current-carrying wire with your right hand; the fingers close about it and the thumb points along the wire in the direction in which the current is flowing. If you do that, then the sweep of the curving fingers, from palm to fingernails, indicates the direction of the magnetic lines of force.

(It is quite possible to make use of the direction of electron flow instead of that of the conventional current. The electron flow is in the direction opposite to that of the current, so that if you use the same device but imagine yourself to be seizing the wire with the left hand rather than the right, and the thumb in the direction of the electron flow, the fingers will still mark out the lines of force.)

Just as a magnet can come in a variety of shapes and by no means need be restricted in form to that of a simple bar, so a current-carrying wire can come in a variety of shapes. For instance, the wire can be twisted into a loop. When that happens, the lines of force outside the loop are pulled apart, while those inside the loop are crowded together. In other words, the magnetic field is stronger inside the loop than outside.

Now suppose that instead of one loop, the wire is twisted into a number of loops, so that it looks like a bedspring. Such a shape is called a *helix* or *solenoid* (the latter word from a Greek expression meaning "pipe-shaped"). In such a solenoid, the lines of force of each loop would reinforce those of its neighbor in such a way that the net result is a set of lines of force that sweep round the exterior of the entire solenoid from one end to the other. They then enter the interior of the solenoid to return to the first end again. The more individual loops or coils in the solenoid, the greater the reinforcement and the more lines of force are crowded into the interior. If the coils are pushed closed together, the reinforcement is more efficient, and again the magnetic flux increases in the interior of the solenoid.

In other words, the flux within the solenoid varies directly with the number of coils (N) and inversely with the length (L). It is proportional then to N/L. The strength of the magnetic field produced by a flowing current depends also on the current intensity. A 2-ampere current will produce twice the magnetic force at a given distance from the wire that a 1-ampere current will. In the case of the solenoid, then, we have the following relationship for a magnetic field that is virtually uniform in strength through the interior:

$$H = \frac{1.25 \; NI}{L}$$ (Equation 12–1)

where H is the strength of the magnetic field in oersteds, I the current intensity in amperes, N the number of turns in the solenoid, and L the length of the solenoid in centimeters.

The relationship between the strength of the magnetic field and the current intensity makes it possible to define the ampere in terms of the magnetic forces set up. If, in two long straight parallel conductors, one meter apart in air, constant and equal currents are flowing so as to produce a mutual force (either attractive or repulsive) of 2×10^{-7} newtons per meter of length, those currents have an intensity of 1 ampere. In this way, the ampere can be defined on the basis of mechanical units only, and all other electrical units can then be defined in terms of the ampere. (It was because Ampère's work made it possible to supply such a mechanical definition of an electrical unit that his name was given to the unit.)

A solenoid behaves as though it were a bar magnet made out of air. This, of course, suggests that in ordinary bar magnets there is a situation that is analogous to the looping electric current of the solenoid. It was not until the twentieth century, however, that the existence of the electron and its relationship to the atom came to be well enough understood to make that thought more than a vague speculation. Only then could ordinary magnetism be pictured as the result of spinning electron charges within atoms. In some cases, electron spins within atoms can be balanced, as some spin clockwise and some counterclockwise in such a fashion that no net magnetic force is to be observed. In other cases, notably in that of iron, the spins are not balanced and the magnetic force can make itself quite evident if the atoms themselves are properly aligned.

Furthermore, this offers the possibility of explaining the earth's magnetism. Even granted that the earth's liquid iron is at a temperature above the Curie point (see page 148) and cannot be an ordinary magnet, it is nevertheless possible that the earth's rotation sets up slow eddies in that liquid core, that electric charge is carried about in those eddies, and that the earth's core behaves as a solenoid rather than as a bar magnet. The effect would be the same.

If this is so, a planet that does not have a liquid core that can carry eddies, or that does not rotate rapidly enough to set eddies into motion, ought not to have much, if any, magnetic field. So far, the facts gathered by contemporary rocket experiments seem to bear this out. The density of the moon is only three-fifths that of the earth, which makes it seem likely that the moon does not have any high-density liquid iron core of significant size; and lunar probes have made it plain that the moon has, indeed, no significant magnetic field.

Venus, on the other hand, is very like the earth in size and density and therefore is quite likely to have a liquid iron core. However, astronomical data gained in the 1960's make it appear likely that Venus rotates slowly indeed, perhaps only once in 200-plus days. And Venus, too, according to observations of the Mariner II Venus-probe, lacks a significant magnetic field.

Jupiter and Saturn, which are much larger than the earth, nevertheless rotate more rapidly and possess magnetic fields far more intense than that of the earth.

The sun itself is fluid throughout, though gaseous rather than liquid, and as the result of its rotation, eddies are undoubtedly set up. It is possible that such eddies account for the magnetic field of the sun, which makes itself most evident in connection with the sunspots. Stars have been located with magnetic fields much more intense than that of the sun, and the galaxies themselves are thought to have galaxy-wide magnetic fields.

Applications of Electromagnetism

The strength of the magnetic field within a solenoid can be increased still further by inserting a bar of iron into it. The high permeability of iron (see page 156) will concentrate the already crowded lines of force even more. The first to attempt this was an English experimenter, William Sturgeon (1783–1850), who in 1823 wrapped eighteen turns of bare copper wire about a U-shaped bar and produced an *electromagnet*. With the current on, Sturgeon's electromagnet could lift twenty times its own weight of iron. With the current off, it was no longer a magnet and would lift nothing.

The electromagnet came into its own, however, with the work of the American physicist Joseph Henry (1797–1878). In 1829, he repeated Sturgeon's work, using insulated wire. Once the wire was insulated it was possible to wind it as closely as possible without fear of short circuits from one wire across to the next. Henry could therefore use hundreds of turns in a short length, thus vastly increasing the ratio N/L (see Equation 12–1) and correspondingly increasing the strength of the magnetic field for a given current intensity. By 1831, he had produced an electromagnet, of no great size, that could lift over a ton of iron.

In fact, electromagnetism made possible the production of magnetic fields of unprecedented strength. A toy horseshoe magnet can produce a magnetic field of a few hundred gauss in strength, a good bar magnet can produce one of 3000 gauss, and an excellent one a field of 10,000 gauss. With an electromagnet, however, fields of 60,000 gauss are easily obtainable.

To go still higher is in theory no problem, since one need only increase the current intensity. Unfortunately this also increases the heat produced (heat increases as the square of the current intensity), so the problem of cooling the coils soon becomes an extremely critical one. In addition, the magnetic forces set up large mechanical strains. By the twentieth century, ingenious designing and the use of strong materials had made possible the production of temporary fields in the hundreds of thousands of gauss by means of briefly pulsing electric currents. There was even the momentary production of fields of a million and a half gauss while the conducting material was exploding.

Such intense magnetic fields require the continuing use of enormous electric currents and lavish cooling setups. They are therefore exorbitantly expensive. The possibility arose of avoiding much of this expense by taking advantage of the phenomenon of superconductivity (see page 188). If certain conductors are cooled to liquid helium temperatures, their resistance drops to zero, so that current flowing through them develops no heat at all no matter how high the current intensity. Furthermore, an electric current set up in a closed circuit at such temperatures continues flowing forever, and a magnetic field set up in association with it also maintains itself forever (or at least as long as the temperature is maintained sufficiently low). In other words, the magnetic field need not be maintained at the expense of a continuous input of current.

If a superconductive material is used as the windings about an iron core and the whole is kept at liquid helium temperatures, then it might seem that by pumping more and more electricity into it, higher and higher magnetic field strengths could be built up without limit, and that when a desired level is reached, current can be shut off and the field left there permanently.

Unfortunately, superconductivity will not quite open the door in that fashion. A superconductive material is perfectly diamagnetic—that is, no magnetic lines of force at all will enter the interior of the superconductive material. The two properties —superconductivity and perfect diamagnetism—are bound together. If more and more current is pumped into a superconductive electromagnet and the magnetic field strength is built higher and higher, the magnetic flux mounts. The lines of force crowd more and more densely together, and at some point (the *critical field strength*) they are forced into the interior of the superconductive material. As soon as the material loses its perfect diamagnetism, it also loses its superconductivity; heat development starts, and the whole process fails. A superconductive magnet cannot be stronger than the critical field strength of the material making up the coils, and unfortunately, this is in the mere hundreds of gauss for most metals. Lead, for instance, will lose its superconductivity at even the lowest possible temperatures in a magnetic field of 600 gauss. Superconductive magnets of lead can therefore be built no stronger than a toy.

Fortunately, it was discovered in the 1950's that much more could be done with alloys than with pure metals. For instance, an alloy of niobium and tin can maintain superconductivity at liquid helium temperatures even while carrying enough current to produce a continuous and relatively cheap magnetic field in excess of 200,000 gauss, while an alloy of vanadium and gallium may do several times as well. The age of high-intensity superconducting electromagnets would seem to be upon us.

The electromagnet is fit for more than feats of brute strength, however. Consider a circuit that includes an electromagnet. There is a key in the circuit that is ordinarily, through some spring action, kept in an open position, so that an air gap is continuously present and no current flows through the circuit. If the key is depressed by hand so that the circuit is closed, the current flows and the electromagnet exerts an attracting force upon a bar of iron near it.

However, suppose this bar of iron is itself part of the circuit and that when it is attracted to the electromagnet it is pulled away from a connection it makes with the remainder of the circuit. The circuit is broken and the current stops. As soon as the current stops, however, the electromagnet is no longer exerting an attracting force, and the iron bar is pulled back to its original position by an attached spring. Since this closes the circuit again, the electromagnet snaps into action and pulls the iron bar toward itself again.

As long as the key remains depressed, this alternation of the iron bar being pulled to the magnet and snapping back to the circuit will continue. In itself this will make a rapid buzzing sound, and the result is, indeed, what is commonly called a "buzzer." If a clapper that strikes a hemisphere of metal is attached to the iron bar, we have an *electric bell*.

But suppose the iron bar is not itself part of the circuit. In that case, when the key is depressed the electromagnet gains attracting power and pulls the bar to itself and keeps it there. Once the key is released (and not until then), the electromagnet loses its attracting force and the bar snaps back.

The iron bar, under these circumstances, snaps back and forth, not in one unvarying rapid pattern of oscillation, but in whatever pattern one cares to impose upon the key as it is depressed and released. The iron bar makes a clicking sound as it strikes the magnet, and the pattern of the hand's movement upon the key is transformed into a pattern of clicks at the electromagnet.

It might occur to anyone that this could be used as the basis of a code. By letting a particular click pattern stand for particular letters of the alphabet, a message could be sent from one place to another at near the speed of light.

The practical catch is that the current intensity that can be pushed through a wire under a given potential difference decreases as the wire lengthens and its total resistance increases. Over really long distances, the current intensity declines to insignificance, unless prohibitive potential differences are involved, and is then not sufficient to produce a magnetic field strong enough to do the work of moving the heavy iron bar.

It was Henry who found a way to solve this problem. He ran current through a long wire until it was feeble indeed, but still strong enough to activate an electromagnet to the point where a lightweight key could be pulled toward it. This lightweight key, in moving toward the electromagnet, closed a second circuit that was powered by a battery reasonably near the key, so that a current was sent through a second, shorter wire. Thanks to the shortness of the second wire and its consequently smaller resistance, this second current was much stronger than the first. However, this second, stronger current mirrored the first, weaker current exactly, for when the original key was depressed by hand, the far-distant key was at once depressed by the electromagnet, and when the original key was released by hand, the far-distant key was at once released by the electromagnet.

Such a device, which passes on the pattern of a current from one circuit to another, is an *electric relay*. The second circuit may be, in its turn, a long one that carries just enough current intensity to activate a third circuit that, in turn, can just activate a far-distant fourth circuit. By using relays and batteries at regular intervals, there is nothing, in principle, to prevent one from sending a particular pattern of clicks around the world. By 1831, Henry was sending signals across a mile of wire.

Henry did not attempt to patent this or to develop it into a practical device. Instead, he helped an American artist, Samuel Finley Breese Morse (1791–1872) do so. By 1844, wires had been strung from Baltimore to Washington, and a pattern of clicks (reproduced as dots for short clicks and dashes for long ones—the "Morse code") was sent over it. The message was a quotation from the Bible's book of Numbers: "What hath God wrought?" This marks the invention of the *telegraph* (from Greek words meaning "writing at a distance"), and the general public was for the first time made aware of how the new electrical science could be applied in a manner that would change man's way of life.

Eventually telegraph lines spanned continents, and by 1866 a cable had been laid across the Atlantic Ocean. Through the cable, the Morse code could pass messages between Great Britain and the United States almost instantly. Laying the cable was a difficult and heartbreaking task, carried through only by the inhuman perseverance of the American financier Cyrus West Field (1819–1892). Its operation was also attended by enormous difficulties, since relays could not be set up at the bottom of the sea as they could be on land. Many problems had to be solved by such men as the British physicist William Thomson, Lord Kelvin (1824–1907), and even so, intercontinental communication did not become quite satisfactory until the invention of the radio (a matter which will be discussed in Volume III of this book). Nevertheless, by 1900 no civilized spot on earth was out of reach of the telegraph, and after thousands of years of civilization, mankind was for the first time capable of forming a single (if not always a mutually friendly or even mutually tolerant) community.

A more direct method of communication also depends in great part on the electromagnet. This is the *telephone* ("to speak at a distance"), invented in 1876 by a Scottish-American speech teacher, Alexander Graham Bell (1847–1922), and shortly afterward improved by Edison.

To put it as simply as possible, the telephone transmitter (into which one speaks) contains carbon granules in a box bounded front and back by a conducting wall. The front wall is a rather thin and, therefore, flexible diaphragm. Through this box an electric current flows. The resistance of the carbon granules depends on how well they make contact with each other. The better the contact, the lower the overall resistance and (since the potential difference remains constant) the greater the current intensity flowing through it.

As one speaks into the transmitter, sound waves set up a complex pattern of compressions and rarefactions in the air (see page I–156). If a region of compression strikes the diaphragm making up the front end of the box of carbon granules, the diaphragm is pushed inward. When a region of rarefaction strikes it, it is pulled outward. It acts precisely as does the eardrum, and in its motion, mimics all the variations in the compression and rarefaction of the sound wave pattern.

When the diaphragm is pushed inward, the carbon granules make better contact, and the current intensity rises in proportion to the extent to which the diaphragm is pushed. Similarly, the current intensity falls as the diaphragm is pulled outward, so that carbon granules fall apart and make poorer contact. Thus, an electric current is set up in which the intensity varies in precise imitation of the compression-rarefaction pattern of the sound wave.

At the other end of the circuit (which may be, thanks to relays and other refinements, thousands of miles away) the current activates an electromagnet in a telephone receiver. The strength of the magnetic field produced varies with the current intensity, so the strength of this field precisely mimics the sound wave pattern impinging upon the far-off transmitter. Near the electromagnet is a thin iron diaphragm that is pulled inward by the magnetic force in proportion to the strength of that force. The diaphragm in the receiver moves in a pattern that mimics the sound wave pattern impinging upon the transmitter many miles away and, in its turn, sets up a precisely similar sound wave pattern in the air adjoining it. The result is that the ear at the receiver hears exactly what the mouth at the transmitter is saying.

Examples of newer applications of electromagnets involve superconductivity (see page 156). A disk which is itself superconductive might rest above a superconducting magnet. The magnetic lines of force will not enter the perfectly diamagnetic superconducting disk, which cannot, for that reason, maintain physical contact with the magnet. There must be room between the two superconducting materials to allow passage, so to speak, for the lines of force. The disk, therefore, is repelled by the magnet and floats above it. Even weights placed upon the disk will not (up to some certain limit) force it down into contact with the magnet. Under conditions worked with in the laboratory, disks capable of bearing weights up to 300 grams per square centimeter have been demonstrated. Without physical contact, the disk can rotate virtually frictionlessly and thus may serve as a frictionless bearing.

Tiny switches can be made by taking advantage of electromagnetism under superconductive conditions. The first such device to be developed (as long ago as 1935) consisted of a thin wire of niobium about a thicker wire of tantalum. Both are superconducting materials, but they can be arranged to have different critical field strengths. A small current can be set up within the tantalum, for instance, and this will maintain itself indefinitely as long as the temperature is kept low. If an even smaller current is sent through the niobium coil about the tantalum, however, the magnetic field set up is sufficient to destroy the superconductivity of the tantalum (while not affecting that of the niobium). The current in the tantalum therefore ceases. In this way, one current can be switched off by another current.

Such a device is called a *cryotron* (from a Greek word meaning "to freeze," in reference to the extremely low temperatures required for superconductivity to evidence itself). Complex combinations of cryotrons have been made use of as versatile switching devices in computers. The advantage of cryotron switches are that they are very fast, very small, and consume very little energy. There is the disadvantage, of course, that they will only work at liquid helium temperatures.

Measurement of Current

The electromagnet introduced a new precision into the study of electricity itself. It made it possible to detect currents by the presence of the magnetic field they created and to estimate the current intensity by the strength of the magnetic field.

In 1820, following hard upon Oersted's announcement of the magnetic field that accompanied a flowing current, the German physicist Johann Salomo Christoph Schweigger (1779–1857) put that field to use as a measuring device. He placed a magnetized needle within a couple of loops of wire. When current flowed in one direction, the needle was deflected to the right; when current flowed in the other direction, the needle was deflected to the left. By placing a scale behind the needle, he could read off the amount of deflection and therefore estimate the current intensity. This was the first *galvanometer* ("to measure galvanic electricity"), a name suggested by Ampère.

Schweigger's original galvanometer had a fixed coil of wire and a movable magnet, but with time it was found more convenient to have a fixed magnet and a movable coil. The device still depends on the deflection of a needle, but now the needle is attached to the coil rather than to the magnet. A particularly practical device of this type was constructed in 1880 by the French physicist Jacques Arsène d'Arsonval (1851–1940) and is known as a *D'Arsonval galvanometer.*

Galvanometers can be made sensitive enough to record extremely feeble current intensities. In 1903, the Dutch physiologist Willem Einthoven (1860–1927) invented a *string galvanometer.* This consisted of a very fine conducting fiber suspended in a magnetic field. Tiny currents flowing through the fiber would cause its deflection, and by means of such an extremely sensitive galvanometer, the small changes in current intensities set up in a contracting muscle could be detected and measured. In this way, the shifting electric pattern involved in the heartbeat could be studied, and an important diagnostic device was added to the armory of modern medicine.

Galvanometers, in fact, tend to be so sensitive that in unmodified form they may safely be used only for comparatively feeble current intensities. To measure the full intensity of an ordinary household current, for instance, the galvanometer must be deliberately short-circuited. Instead of allowing all the current to flow through the moving coil in the galvanometer, a low-resistance conductor is placed across the wires leading in and out of the coil. This low-resistance short circuit is called a shunt. The device was first used in 1843 by the English physicist Charles Wheatstone (1802–1875).

Shunt and coil are in parallel and the current intensity going through each is in inverse proportion to their respective resistances. If the resistances are known, we can calculate what fraction of the current intensity will travel through the coil, and it is that fraction only that will influence the needle deflection. The sensitivity of the deflection can be altered by adding or subtracting resistances to the shunt, thus decreasing or increasing the fraction of the total current intensity passing through the coil.

By adjusting the fraction of the current intensity reaching the coil so that the deflected needle will remain on the scale, household current intensities or, in principle, current of any intensity, can be measured. The dial can be calibrated to read directly in amperes, and a galvanometer so calibrated is called an *ammeter* (a brief form of "ampere-meter").

Suppose a galvanometer is connected across some portion of a circuit, short-circuiting it. If this galvanometer includes a very high resistance, however, a current of very little intensity will flow over the short circuit through the galvanometer—current of an intensity low enough not to affect the remainder of the circuit in any significant way.

This small current intensity will be under the driving force of the same potential difference as will the current flowing in much greater intensity through the regular route of the circuit between the points across which the galvanometer has been placed. The tiny current intensity driven through the high-resistance galvanometer will vary with the potential difference. The scale behind the moving needle can then be calibrated in volts, and the galvanometer becomes a *voltmeter*.

Once current intensity and potential difference through some circuit or portion of a circuit is measured by an ammeter and voltmeter, the resistance of that same circuit or portion of a circuit can be calculated by Ohm's law. However, with the help of a galvanometer, resistance can also be measured directly by balancing the unknown resistance against known resistances.

Suppose a current is flowing through four resistances—R_1, R_2, R_3, and R_4—arranged in a parallelogram. Current enters at A and can flow either through B to D via R_1 and R_2, or through C to D via R_3 and R_4. Suppose that a conducting wire connects B and C and that a galvanometer is included as part of that wire. If the current reaches B at a higher potential than it reaches C, current will flow from B to C and the galvanometer will register in one direction. If the reverse is true and the current reaches C at a higher potential than at B, current will flow from C to B and the galvanometer will register in the other direction But if the potential at B and the potential at C are exactly equal, current will flow in neither direction and the galvanometer will register zero.

Suppose the galvanometer does register zero. What can we deduce from that? The current flowing from A to B must pass on, intact, from B to D; none is deflected across the galvanometer. Therefore the current intensity from A to B through R_1 must be the same as the current intensity from B to D through R_2. Both intensities can be represented as I_1. By a similar argument, the current intensities passing through R_3 and R_4 are equal and may be symbolized as I_2.

By Ohm's law, potential difference is equal to current intensity times resistance $(E = IR)$. The potential difference from A to B is therefore $I_1 R_1$; from B to D is $I_1 R_2$; from A to C is $I_2 R_3$; and from C to D is $I_2 R_4$.

But if the galvanometer reads zero, then the potential difference from A to B is the same as from A to C (or current would flow between B and C and the galvanometer would not read zero), and the potential difference from B to D is the same as from C to D, by the same argument. In terms of current intensities and resistances, we can express the equalities in potential difference thus:

$$I_1 R_1 = I_2 R_3 \qquad \text{(Equation 12–2)}$$

$$I_1 R_2 = I_2 R_4 \qquad \text{(Equation 12–3)}$$

If we divide Equation 12–2 by Equation 12–3, we get:

$$\frac{R_1}{R_2} = \frac{R_3}{R_4} \qquad \text{(Equation 12–4)}$$

Now suppose that R_1 is the unknown resistance that we want to measure, while R_2 is a known resistance. As for R_3 and R_4, they are *variable resistances* that can be varied through known steps.

A very simple variable resistance setup can consist of a length of wire stretched over a meter stick, with a sliding contact capable of moving along it. The sliding contact can represent point C in the device described above. The stretch of wire from end to end of the meter stick is AD. That portion stretching from A to C is R_3 and the portion from C to D is R_4. If the wire is uniform, it is fair to assume that the resistances R_3 and R_4 will be in proportion to the length of the wire from A to C and from C to D respectively, and those lengths can be read directly off the meter stick. The absolute values of R_3 and R_4 cannot be determined, but the ratio R_3/R_4 is equal to AC/CD and that is all we need.

As the sliding contact is moved along the wire, the potential difference between A and C increases as the distance between the two points increases. At some point the potential difference between A and C will become equal to that between A and D, and the galvanometer will indicate that point by registering zero. At that point the ratio R_3/R_4 can be determined directly from the meter stick, and the ratio R_1/R_2 can, by Equation 12–4, be taken to have the same value.

The unknown resistance, R_1, can then easily be determined by multiplying the known resistance R_2 by the known ratio R_3/R_4. Wheatstone used this device to measure resistances in 1843, and (athough some researchers had used similar instruments before him) it has been called the *Wheatstone bridge* ever since.

Generators

The electromagnet, however useful, does not in itself solve the problem of finding a cheap source of electricity. If the magnetic field must be set up through the action of a chemical cell, the field will remain as expensive as the electric current that sets it up, and large-scale uses will be out of the question.

However, the manner in which an electromagnet was formed was bound to raise the question of the possibility of the reverse phenomenon. If an electric current produces a magnetic field, might not a magnetic field already in existence be used to set up an electric current?

Michael Faraday thought so, and in 1831 he attempted a crucial experiment (after having tried and failed four times before). In this fifth attempt, he wound a coil of wire around one segment of an iron ring, added a key with which to open and close the circuit, and attached a battery. Now when he pressed the key and closed the circuit, an electric current would flow through the coil and set up a magnetic field. The magnetic lines of force would be concentrated in the highly permeable iron ring in the usual fashion.

Next, he wound another coil of wire about the opposite segment of the iron ring and connected that coil to a galvanometer. When he set up the magnetic field, it might start a current flowing in the second wire and that current, if present, would be detected by the galvanometer.

The experiment did not work as he expected it to. When he closed the circuit there was a momentary surge of current through the second wire, as was indicated by a quick deflection of the galvanometer needle followed by a return to a zero reading. The zero reading was then maintained however long the key remained depressed. There was a magnetic field in existence and it was concentrated in the iron ring, as could easily be demonstrated. However, the mere existence of the magnetic field did not in itself produce a current. Yet when Faraday opened the circuit again, there was a second quick deflection of the galvanometer needle, in a direction opposite to the first.

Faraday decided it was not the existence of magnetic lines of force that produced a current, but the motion of those lines of force across the wire. He pictured matters after this fashion. When the current started in the first coil of wire, the magnetic field sprang into being, the lines of force expanding outward to fill space. As they cut across the wire in the second coil, a current was initiated. Because the lines of force quickly expanded to the full and then stopped cutting across the wire, the current was only momentary. With the circuit closed and the magnetic field stationary, no further electric current in the second coil was to be expected. However, when he opened the first circuit, the magnetic field ceased and the lines of force collapsed inward again, momentarily setting up a current in a direction opposite to the first.

He showed this fact more plainly to himself (and to the audiences to whom he lectured) by inserting a magnet into a coil of wire that was attached to a galvanometer. While the magnet was being inserted, the galvanometer needle kicked in one direction; and while it was being withdrawn, it kicked in the other direction. While the magnet remained at rest within the coil at any stage of its insertion or withdrawal, there was no current in the coil. However, the current was also initiated in the coil if the magnet was held stationary and the coil was moved down over it or lifted off again. It didn't matter, after all, whether the wire moved across the lines of force, or the lines of force moved across the wire.*

Faraday had indeed used magnetism to induce an electric current and had thus discovered *electromagnetic induction*. In the United States, Henry had made a similar discovery at about the same time, but Faraday's work was published first.

The production of an induced current is most easily visualized if one considers the space between the poles of a magnet, where lines of force move across the gap in straight lines from the north pole to the south pole, and imagines a single copper wire moving between those poles. (It makes no difference, by the way, whether the magnet in question is a permanent one or an electromagnet with the current on.)

If the wire were motionless or were moving parallel to the lines of force, there would be no induced current. If the wire moved in a direction that was not parallel to the lines of force, so that it cut across them, then there would be an induced current.

The size of the potential difference driving the induced current would depend upon the number of lines of force cut across per second, and this in turn would depend on a number of factors. First there is the velocity of the moving wire. The more rapidly the wire moves in any given direction not parallel to the lines of force, the greater the number of lines of force cut across per second, and the greater the potential difference driving the induced current.

Again, there is the question of the direction of the motion of the wire. If the wire is moving in a direction perpendicular to the lines of force, then it is cutting across a certain number of lines of force per second. If the wire is moving, at the same speed, in a direction not quite perpendicular to the lines of force, it cuts across fewer of them per unit time, and the potential difference of the induced current is less intense. The greater the angle between the direction of motion and the perpendicular to the lines of force, the smaller the potential difference of the induced current. Finally when the motion is in a direction 90° to the perpendicular, that

* There is a famous story about such a demonstration, which is probably apocryphal. A woman, having watched the demonstration of the coil and the magnet, is supposed to have said, "But, Mr. Faraday, of what use is this?" Faraday is reported to have answered politely, "Madam, of what use is a newborn baby?" Another version has William Ewart Gladstone (then a freshly elected member of Parliament and eventually to be four times Prime Minister) ask the question. Faraday is supposed to have answered, "Sir, in twenty years, you will be taxing it."

motion is actually parallel to the lines of force and there is no induced current at all.

In addition, if the wire is in coils, and each coil cuts across the lines of force, the potential difference driving the induced current is multiplied in intensity by the number of coils per unit length.

The direction of the induced current can be determined by using the right hand, according to a system first suggested by the English electrical engineer John Ambrose Fleming (1849–1945) and therefore called *Fleming's rule*. It is applied without complication when the wire is moving in a direction perpendicular to that of the lines of force. To apply the rule, extend your thumb, index finger and middle finger so that each forms a right angle to the other two—that is, allowing the thumb to point upward, the forefinger forward, and the middle finger leftward. If, then, the forefinger is taken as pointing out the direction of the magnetic lines of force from north pole to south pole, and the thumb as pointing out the direction in which the wire moves, the middle finger will point out the direction (from positive pole to negative pole) of the induced current in the wire.

Two months after his discovery of electromagnetic induction, Faraday took his next step. Since an electric current was produced when magnetic lines of force cut across an electrical conductor, how could he devise a method of cutting such lines continuously?

He set up a thin copper disk that could be turned on a shaft. Its outer rim passed between the poles of a strong magnet as the disk turned. As it passed between those poles, it continuously cut through magnetic lines of force, so that a potential difference was set up in the disk, a difference that maintained itself as long as the disk turned. Two wires ending in sliding contacts were attached to the disk. One contact brushed against the copper wheel as it turned, the other brushed against the shaft. The other ends of the wires were connected to a galvanometer.

Since the electric potential was highest at the rim where the material of the disk moved most rapidly and therefore cut across more lines of force per unit time, a maximum potential difference existed between that rim and the motionless shaft. An electric current flowed through the wires and galvanometer as long as the disk turned. Faraday was generating a current continuously, without benefit of chemical reactions, and had thus built the first *electric generator*.

The importance of the device was tremendous, for, in essence, it converted the energy of motion into electrical energy. A disk could be kept moving by a steam engine, for instance, at the expense of burning coal or oil (much cheaper than burning zinc), or by a turbine that could be turned by running water, so that streams and waterfalls could be made to produce electricity. It took fifty years to work out all the technical details that stood in the way of making the generator truly practical, but by the 1880's cheap electricity in quantity was a reality; and the electric light and indeed the electrification of society in general became possible.

13

Alternating Current

In modern generators, Faraday's copper disk turning between the poles of a magnet is replaced by coils of copper wire wound on an iron drum turning between the poles of an electromagnet. The turning coils make up the *armature*. To see what happens in this case, let's simplify matters as far as possible and consider a single rectangular loop of wire rotating between a north pole on the right and a south pole on the left.

Imagine such a rectangle oriented parallel to the lines of force (moving from right to left) and beginning to turn in such a fashion that the wire on the left side of the rectangle (the L wire) moves upward, across the lines of force, while the wire on the right side of the rectangle (the R wire) moves downward, across the lines of force.

Concentrate, to begin with, on the L wire and use the Fleming right-hand rule. Point your thumb upward, for that is the direction in which the L wire is moving. Point your forefinger left, for that is the direction of the magnet's south pole. Your middle finger points toward you, and that is the direction of the induced current in the L wire.

What about the R wire? Now the thumb must be pointed downward while the forefinger still points left. The middle finger points away from you and that is the direction of the induced current in the R wire. If the induced current is traveling toward you in the L wire and away from you in the R wire, you can see that it is actually going round and round the loop.

Next imagine that the L wire and R wire are both connected to separate "slip rings" (Ring A and Ring B, respectively), each of which is centered about the shaft that serves as an axis around which the loop rotates. The current would tend to flow from Ring B through the R wire into the L wire and back into Ring A. If one end of a circuit is connected to one ring by way of a brushing contact, and the other end of the same circuit to the other ring, the current generated in the turning armature would travel through the entire circuit.

But let's consider the rectangular loop a bit longer. Since the loop is rotating, the L wire and R wire cannot move up and down, respectively, indefinitely. They are, in fact, constantly changing direction. As the L wire moves up, it curves to the right and moves at a smaller angle to the lines of force, so that the intensity of the induced current decreases. Precisely the same happens to the R wire, for as it moves downward, it curves to the left and moves also at a smaller angle to the lines of force.

The current continues to decrease as the loop turns until the loop has completed a right angle turn, so that the L wire is on top and the R wire on bottom. The L wire is now moving right, parallel to the lines of force, while the R wire is moving left, also parallel to the lines of force. The intensity of the induced current has declined to zero. As the loop continues to rotate, the L wire cuts down into the lines of force, while the R wire cuts up into them. The two wires have now changed places, the L wire becoming the R wire, and the R wire becoming the L wire.

The wires, despite this change of place (as far as the direction of the induced current is concerned), are still connected to the same slip rings. This means that as the armature makes one complete rotation, the current flows from Ring B to Ring A for half the time and from Ring A to Ring B for the other half. This repeats itself in the next rotation, and the next, and the next. Current produced in this manner is therefore *alternating current* (usually abbreviated *AC*) and moves backward and forward perpetually. One rotation of the loop produces one back and forth movement of the current—that is, one *cycle*. If the loop rotates sixty times a second we have a 60-cycle alternating current.

Nor is the current steady in intensity during the period when it is moving in one particular direction. During one rotation of the loop, the current intensity begins at zero, when the moving wires (top and bottom) are moving parallel to the lines of force; it rises smoothly to a maximum, when the wires (right and left) are moving perpendicular to the lines of force, and then drops just as smoothly to zero again, when the wires (bottom and top) are moving parallel to the lines of force once more.

As the loop continues turning, the current changes direction, and we can now imagine the flow to be less than zero—that is, we can decide to let the current intensity be measured in positive numbers when its flow is in one direction and in negative numbers when its flow is in the other. Therefore, after the intensity has dropped to zero, it continues smoothly dropping to a minimum when the wires (left and right) are moving perpendicular to the lines of force; and it rises smoothly to zero again when the wires (top and bottom) are moving parallel to the lines of force once more. This completes one rotation, and the cycle begins again.

If, for convenience, we imagine the maximum current intensity to be 1 ampere, then in the first quarter of the rotation of the loop the intensity changes from 0 to $+1$; in the second quarter from $+1$ to 0; in the third quarter from 0 to -1; and in the fourth quarter from -1 to 0. If this change in intensity is plotted against time, there is a smoothly rising and falling wave, endlessly repeated, which mathematicians call a *sine curve*.

A generator can easily be modified in such a way as to make it produce a current that flows through a circuit in one direction only. This would be a *direct current* (usually abbreviated *DC*), and it is this type of current that was first dealt with by Volta and which is always produced by chemical cells.

Suppose the two ends of our rectangular loop are attached to "half-rings" which adjoin each other around the shaft serving as axis of rotation but which don't touch. The L wire is attached to one half-ring and the R wire to the other. The brush contact of one end of the circuit touches one half-ring; the brush contact of the other touches the second half-ring.

During the first half of the rotation of the armature, the current flows, let us say, from Half-Ring A to Half-Ring B. During the second half, the current reverses itself and flows from Half-Ring B to Half-Ring A. However, every time the armature goes through a half-rotation, the half-rings change places. If one brush contact is touching the positive half-ring, the negative half-ring turns into place just as it becomes positive, and leaves its place just as it begins to turn negative again. In other words, the first brush contact is touching each half-ring in turn, always when the rings are in the positive portion of their cycle; the other brush contact always touches the half-rings when they are negative. The current may change direction in the armature, but it flows in one constant direction in the attached circuit.

The intensity still rises and falls, to be sure, from 0 to +1 to 0 to +1, and so on. By increasing the number of loops and splitting the rings into smaller segments, these variations in intensity can be minimized, and a reasonably constant direct current can be produced.

The AC generator is simpler in design than the DC generator, but alternating current had to overcome a number of objections before it could be generally accepted. Edison, for instance, was a great proponent of direct current and, during the closing decades of the nineteenth century, fought the use of alternating current bitterly. (The great proponent of alternating current was the American inventor George Westinghouse [1846–1914]).

In considering this competition between the two types of current, it may seem natural at first that DC should be favored. In fact, AC may seem useless on the face of it. After all, a direct current is "getting somewhere" and is therefore useful, while an alternating current "isn't getting anywhere" and therefore can't be useful—or so it might seem.

Yet the "isn't getting anywhere" feeling is wrong. It is perhaps the result of a false analogy with water running through a pipe. We usually want the water to get somewhere—to run out of the pipe, for instance, so that we can use it for drinking, washing, cooling, irrigating, fire fighting, and so on.

But electricity never flows out of a wire in ordinary electrical appliances. It "isn't getting anywhere" under any circumstances. Direct current may go in one direction only, but it goes round and round in a circle and that is no more "getting anywhere" than moving backward and forward in one place.

There are times when DC is indeed necessary. In charging storage batteries, for instance, you want the current to move always in the direction opposite to that in which the current moves when the storage battery is discharging. On the other hand, there are times when it doesn't matter whether current is direct or alternating. For instance, a toaster or an incandescent bulb works as it does simply because current forcing its way through a resistance heats up that portion of the circuit (to a red-heat in the toaster and to a white-heat in the bulb). The heating effect does not depend upon the direction in which the current is flowing, or even on whether it continually changes direction. By analogy, you can grow heated and sweaty if you run a mile in a straight line, or on a small circular track, or backward and forward in a living room. The heating effect doesn't depend on "getting anywhere."

A more serious objection to AC, however, was that the mathematical analysis of its behavior was more complicated than was that of DC circuits. For the proper design of AC circuits, this mathematical analysis first had to be made and understood. Until then, the circuits were continuously plagued by lowered efficiency.

Impedance

A situation in which the current intensity and the potential difference are changing constantly raises important questions—for instance, as to how to make even the simplest calculations involving alternating current. When a formula includes I (current intensity) or E (potential difference), one is entitled to ask what value to insert when an alternating current has no set value for either quantity, but a value that constantly varies from zero to some maximum value (I_{max} and E_{max}), first in one direction and then in the other.

One must then judge these properties of the alternating current by the effects they produce, rather than by their sheer numerical values. It can be shown, for instance, that an alternating current can, in heat production and in other uses, have the same effect as a direct current with definite values of I and E. The values of I and E therefore represent the *effective current intensity* and the *effective potential difference* of the alternating current, and it is these effective values that I and E symbolize in alternating currents. The effective values are related to the maximum values as follows:

$$I = \frac{I_{max}}{\sqrt{2}} = 0.7 \, I_{max} \qquad \text{(Equation 13–1)}$$

$$E = \frac{E_{max}}{\sqrt{2}} = 0.7 \, E_{max} \qquad \text{(Equation 13–2)}$$

One might suppose that having defined I and E for alternating currents one could proceed at once to resistance, representing that by the ratio of E/I (the current intensity produced by a given potential difference) in accordance with Ohm's law. Here, however, a complication arises. A circuit that under direct current would have a low resistance, as indicated by the fact that a high current intensity would be produced by a given potential difference, would under alternating current have a much greater resistance, as indicated by the low current intensity produced by the same potential difference. Apparently, under alternating current a circuit possesses some resisting factor other than the ordinary resistance of the materials making up the circuit.

To see why this is so, let's go back to Faraday's first experiments on electromagnetic induction (see page 216). There, as the electric current was initiated in one coil, a magnetic field was produced and the expanding lines of force cut across a second coil, inducing a potential difference and, therefore, an electric current in a particular direction in that second coil. When the electric current was stopped in the first coil, the collapsing lines of force of the dying magnetic field cut across the second coil again, inducing a potential difference of reversed sign and, therefore, an electric current in the opposite direction in that second coil.

So far, so good. However, it must be noted that when current starts flowing in a coil, so that magnetic lines of force spread outward, they not only cut across other coils in the neighborhood but also cut across the very coils that initiate the magnetic field. (The lines of force spreading out from one loop in the coil cuts through all its neighbors.) Again, when the current in a coil is cut off, the lines of force of the collapsing magnetic field cut across the very coils in which the current has been cut off. As current starts or stops in the coil, an induced current is set up in that same coil. This is called *self-induction* or *inductance*, and it was discovered by Henry in 1832. (Here Henry announced his discovery just ahead of Faraday, who made the same discovery independently; Faraday, you will remember, just anticipated Henry in connection with electromagnetic induction itself.)

Almost simultaneously with Henry and Faraday, a Russian physicist, Heinrich Friedrich Emil Lenz (1804–1865), studied inductance. He was the first to make the important generalization that induced potential difference set up in a circuit always acts to oppose the change that produced it. This is called *Lenz's law*.

Thus, when a current in a coil is initiated by closing a circuit, one would expect that the current intensity would rise instantly to its expected level. However, as it rises it sets up an induced potential difference, which tends to produce a current in the direction opposed to that which is building up. This opposition by inductance causes the primary current in the circuit to rise to its expected value with comparative slowness.

Again, if the current in a coil is stopped by breaking a circuit, one would expect the current intensity to drop to zero at once. Instead, the breaking of the circuit sets up an induced potential which, again, opposes the change and tends to keep the current flowing. The current intensity therefore drops to zero with comparative slowness. This opposed potential difference produced by self-induction is often referred to as *back-voltage*.

In direct current, this opposing effect of inductance is not terribly important, since it makes itself felt only in the moment of starting and stopping a current, when lines of force are moving outward or inward. As long as the current flows steadily in a single direction, there is no change in the magnetic lines of force, no induced current, and no interference with the primary current itself.

How different for alternating current, however, which is always changing, so that magnetic lines of force are always cutting the coils as they are continually moving either outward or inward. Here an induced potential difference is constantly in being and is constantly opposed to the primary potential difference, reducing its value greatly. Thus, where a given potential difference will drive a high (direct) current intensity through a circuit, under AC conditions it will be largely neutralized by inductance, and will therefore send only a small (alternating) current intensity through the same circuit.

The unit of inductance is the *henry*, in honor of the physicist. When a current intensity in a circuit is changing at the rate of 1 ampere per second and, in the process, induces an opposed potential difference of 1 volt, the circuit is said to have an inductance of 1 henry. By this definition, 1 henry is equal to 1 (volt per ampere) per second, or 1 volt-second per ampere (volt-sec/amp).

The resistance to current flow produced by self-induction depends not only on the value of the inductance itself but also on the frequency of the alternating current, since with increasing frequency, the rate of change in current intensity with time (amperes per second) increases. Therefore, the more cycles per second, the greater the resistance to current flow by a given inductance. Suppose we symbolize inductance as L and the frequency of the alternating current as f. The resistance produced by these two factors is called the *inductive reactance* and is symbolized as X_L. It turns out that:

$$X_L = 2\pi f L \qquad \text{(Equation 13-3)}$$

If *L* is measured in henrys, that is in volt-seconds per ampere, and *f* is measured in per-second units, then the units of X_L must be (volt-seconds per ampere) per second. The seconds cancel and the units become simply volts per ampere, which defines the ohm (see page 185). In other words, the unit of inductive reactance is the ohm, as it is of ordinary resistance.

The ordinary resistance (*R*) and the inductive reactance (X_L) both contribute to the determination of the current intensity placed in an alternating current circuit by a given potential difference. Together they make up the *impedance* (*Z*). It is not a question of merely adding resistance and inductive reactance, however. Impedance is determined by the following equation:

$$Z = \sqrt{R^2 + X_L^2}$$ (Equation 13–4)

In alternating currents, it is impedance that plays the role of ordinary resistance in direct currents. In other words, the AC equivalent of Ohm's law would be $IZ = E$, or $I = E/Z$, or $Z = I/E$.

Reactance is produced in slightly different fashion by condensers. A condenser in a direct current circuit acts as an air gap and, at all reasonable potential differences, prevents a current from flowing. In an alternating current circuit, however, a condenser does not keep a current from flowing. To be sure, the current does not flow across the air gap, but it surges back and forth, piling up electrons first in one plate of the condenser, then in the other. In the passage back and forth from one plate to the other, the current passes through an appliance—let us say an electric light—which proceeds to glow. The filament reacts to the flow of current through itself and not to the fact that there might be another portion of the circuit somewhere else through which there is no flow of current.

The greater the capacitance of a condenser, the more intense the current that sloshes back and forth, because a greater charge can be piled onto first one plate then the other. Another way of putting this is that the greater the capacitance of a condenser, the smaller the opposition to the current flow, since there is more room for the electrons on the plate and therefore less of a pile-up of negative-negative repulsion to oppose a continued flow.

This opposition to a continued flow is the *capacitive reactance* (X_c), which is inversely proportional to the capacitance (*C*) of the condenser. The capacitive reactance is also inversely proportional to the frequency (*f*) of the current, for the more rapidly the current changes direction, the less likely is either plate of the condenser to get an oversupply of electrons during the course of one half-cycle, and the smaller the negative-negative repulsion set up to oppose the current flow. (In other words, raising the frequency lowers the capacitive reactance, though it raises the inductive reactance.) The inverse relationship can be expressed as follows:

$$X_c = \frac{1}{2\pi f C} \qquad \text{(Equation 13–5)}$$

The capacitance (C) is measured in farads—that is in coulombs per volt or, what is equivalent, in ampere-seconds per volt. Since the units of frequency (f) are per-seconds, the units of $2\pi f C$ are ampere-seconds per volt per seconds, or (with seconds canceling) amperes per volt. The units of capacitive reactance (X_c) are the reciprocal of this—that is, volts per ampere, or ohms. Thus capacitive reactance, like inductive reactance, is a form of resistance in the circuit.

Capacitive reactance and inductive reactance both act to reduce the current intensity in an AC circuit under a given potential difference if either one is present singly. However, they do so in opposite manners.

Under simplest circumstances, the current intensity and potential difference of an alternating current both rise and fall in step as they move along the sine curve. Both are zero at the same time; both are at maximum crest or at minimum trough at the same time. An inductive reactance, however, causes the current intensity to *lag;* to reach its maximum (or minimum, or zero point) only a perceptible interval after the potential difference has reached it. A capacitive reactance, on the other hand, causes the current intensity to *lead;* to rise and fall a perceptible period of time ahead of the rise and fall in potential difference. In either case, current intensity and potential difference are out of phase, and energy is lost.

Yet if there is both a capacitive reactance and an inductive reactance in the circuit, the effect of one is to cancel that of the other. The lead of the capacitive reactance must be subtracted from the lag of the inductive reactance. The total impedance can be expressed as follows:

$$Z = \sqrt{R^2 + (X_L - X_c)^2} \qquad \text{(Equation 13–6)}$$

If the circuit is so arranged that the capacitive reactance equals the inductive reactance, then $X_L - X_c = 0$; and $Z =$

$\sqrt{R^2} = R$. The impedance of the alternating current circuit is then no greater than the ordinary resistance of an analogous direct current circuit would be. Such an alternating current circuit is said to be *in resonance*. Notice that the impedance can never be less than the resistance. If the capacitive reactance is greater than the inductive reactance, then $X_L - X_C$ is indeed a negative number, but its square is positive and when that is added to the square of the resistance and the square root of the sum is taken, the final value of Z will be greater than that of R.

This represents only the merest beginnings of the complications of AC circuitry. A good deal of the full treatment was worked out at the beginning of the twentieth century by the German-American electrical engineer Charles Proteus Steinmetz (1865–1923), and it was only thereafter that alternating currents could be properly exploited.

Transformers

Even before Steinmetz had rationalized the use of alternating circuits—and despite the formidable nature of the difficulties which, in the absence of such rationalization, plagued the circuit-designers; and despite, also, the formidable opposition of men such as Edison and Kelvin—the drive for the use of alternating current carried through to victory. The reason for this was that in one respect, that involving the transmission of electric power over long distances, alternating current was supreme over direct current.

The power of an electric current is measured in watts and is equal to the volts of potential difference times the amperes of current intensity. (Strictly speaking, this is true only in the absence of reactance. Where inductive reactance is present, the power is decreased by a particular *power factor*. However, this can be reduced or even eliminated by the addition of a proper capacitive reactance, so this need not bother us.)

This means that different combinations of volts and amperes can represent an electric current of the same power. For instance, a given appliance might carry one ampere at 120 volts, or two amperes at 60 volts, or five amperes at 24 volts, or twelve amperes at 10 volts. In each case, the power would be the same—120 watts.

There are advantages to having a given electric power appear in a high-volt, low-ampere arrangement under some conditions and in a low-volt, high-ampere arrangement under other conditions. In the former case, the low current intensity makes it fear of undue heating effects. In the latter case, the low potential difference means that there is a smaller chance of breaking down the insulation or producing a short circuit.

And then there is the previously mentioned problem of transmitting electric currents over long distances. Much of the convenience of electricity would be lost if it could only be used in the near neighborhood of a generator. Yet if the current is sent through wires over considerable distances, so much energy is likely to be lost in the form of heat that either we have too little electricity at the other end to bother with, or we must reduce the heat-loss by using wire so thick as to be uneconomical. The heat produced, however, is proportional to the square of the current intensity. Therefore, if we reduce the current intensity to a very low quantity, while simultaneously raising the potential difference to a correspondingly high value (in order to keep the total electric power unchanged), much less electricity would be lost as heat.

Naturally, it is not very likely that this arrangement of tiny current intensity combined with huge potential differences would be suitable for use in ordinary appliances. Consequently, we want a situation in which the same power can be at very high voltages for transmission and at very low voltages for use.

With direct current, it is highly impractical to attempt to change the potential difference of a current—now up, now down —to suit changing needs. In alternating current, however, it is easy to do this by means of a *transformer* (a device that "transforms" the volt-ampere relationship). In essence, it was a transformer that Faraday had invented when in 1831 he made use of an iron ring with two sets of wire coils on it in his attempt to induce an electric current.

Faraday found that when a direct electric current was put through one of the coils (the *primary*), no current was induced in the other coil (the *secondary*), except at the moments when the current was initiated or ended. It was only then that magnetic lines of force swept over the secondary.

Where the current in the primary is an alternating current, however, the current intensity is always either rising or falling; and the intensity of the magnetic field through the iron ring is always either rising or falling. The lines of force expand outward and collapse inward over and over, and as they do so, they cut across the secondary, producing an alternating current that keeps in perfect step with the alternating current in the primary.

The potential difference of the induced current depends on

the number of coils in the secondary as compared with the number in the primary. Thus, if the current in the primary has a potential difference of 120 volts and if the secondary contains ten times as many turns of wire as does the primary, then the induced current will have a potential difference of 1200 volts. This is an example of a *step-up transformer*. If the induced current produced by such a transformer is used to power the primary in another transformer in which the secondary now has only one-tenth the number of coils that the primary has, the new induced current is back at 120 volts. The second transformer is a *step-down transformer*.

The induced current (if we ignore negligible losses in the form of heat) must have the same power as the original current. Otherwise, energy will either be created or destroyed in the process, and this is inadmissable. This means that as the potential difference goes up, the current intensity must go down, and vice versa. If a one-ampere current at 120 volts activates a step-up transformer in which the secondary has a hundred times the number of coils that the primary has, the induced current will have a potential difference of 12,000 volts and a current intensity of 1/100 ampere. In both primary and secondary, the power will be 120 watts.

If alternating current generators are used, there is no difficulty at all in altering voltages by means of transformers. A step-up transformer in particular will serve to raise the potential difference to great heights and the current intensity to trivial values. Such a current can be transmitted over long distances through wires that are not excessively thick, with little heat loss, thanks to the low current intensity. Thanks to the high potential difference, however, the full power of the electric current is nevertheless being transmitted.

When the current arrives at the point where it is to be used, a step-down transformer will convert it to a lower potential difference and a higher current intensity for use in household appliances or industrial machines. A particular appliance or machine may need low potential differences at one point and high potential differences at the other, and each can be supplied by the use of appropriate transformers.

Long-distance transmission through high-voltage alternating current was made practical by the work of the Croatian-American electrical engineer Nikola Tesla (1857–1943). He was backed by Westinghouse, who in 1893 won the right to set up at Niagara

Falls a hydroelectric station (where the power of falling water would spin turbines that would turn armatures and produce electricity) for the production and transmission of alternating current.

Since then, AC current has come into virtually universal use, and this is responsible for the great flexibility and versatility of electricity as a form of useful energy.

Motors

Thanks to the development of the generator, mechanical energy could be converted to electrical energy, and it was possible to have large supplies of electricity arising, indirectly, out of burning coal or falling water. Thanks to the development of alternating current and transformers, this electrical energy could be transported over long distances and conducted into every home or factory.

However, once in the home or factory, what was the electricity to do there? Fortunately, by the time electricity could be produced and transported in quantity, the question of the manner of its consumption had already been answered.

That answer arose out of the reversal of a known effect. This happens frequently in science. If deforming the shape of a crystal produces a potential difference, then applying a potential difference to opposite sides of a crystal will deform its shape. If an electric current creates a magnetic field, then a magnetic field can be made to produce an electric current.

It is not surprising, therefore, that if mechanical energy can be converted into electrical energy when a conductor is made to move and cut across lines of magnetic force, electrical energy can be converted into mechanical energy, causing a conductor to move and cut across lines of magnetic force.

Imagine a copper wire between the poles of a magnet, north pole on the right and south pole on the left. If the copper wire is moved upward, then by Fleming's right-hand rule we know that a current will be induced in the direction toward us.

Suppose, however, that we keep the wire motionless in midfield, so that no current is induced in it. Suppose that we then send a current through it from a battery, that current moving toward us. The current-carrying wire now sets up a magnetic field of its own. Since the current is coming toward us, the lines of force run in counterclockwise circles about themselves. Above the wire, those counterclockwise lines of force run in the same direction as do the straight lines of force from the north pole to the south pole of the magnet. The two add together, so that the magnetic flux is increased. Below the wire, the counterclockwise lines of force run in the direction opposed to the lines of force of the magnet, so that there is a canceling of effect and the flux density is decreased. With a high-flux density above the wire, and a low-flux density below it, the wire is pushed downward by the natural tendency of the lines of force to "even out." If the current in the wire is moving away from us, so that its lines of force run in clockwise circles, the magnetic flux density will be greater below and the wire will be pushed upward.

To summarize, consider a magnet with lines of force running from right to left:

If a wire without current is moved upward, current toward you is induced.

If a wire without current is moved downward, current away from you is induced.

If a wire contains current flowing toward you, motion downward is induced.

If a wire contains current flowing away from you, motion upward is induced.

In the first two cases, current is generated out of motion and the device is a generator. In the last two cases, motion is manufactured out of current and the device is a *motor*. (It is the same device in either case, actually, but it is run "forward" in one case and "backward" in the other.)

Notice that in the generator, current toward is associated with motion upward, and current away with motion downward; in the motor, current toward is associated with motion downward, and current away with motion upward. Therefore, in determining the relationship of direction of lines of force, direction of current, and direction of motion, one must, in the case of a motor, use some device that is just the opposite of the device used in the case of the generator.

For the generator we used the Fleming right-hand rule, and since we have an opposite limb in the shape of the left hand, we use a *left-hand rule* (with thumb, forefinger, and middle finger held at mutual right angles) to relate the various directions in a motor. As in the right-hand rule, we allow the forefinger to point in the direction of the lines of force, i.e., toward the south pole. The middle finger points in the direction the current is flowing, i.e., toward the negative pole. The thumb will then automatically point in the direction of the motion imposed upon the wire.

Now let us pass on to a loop of wire between the poles of a

magnet. If a mechanical rotation is imposed upon it, an electric current is induced in the loop. Consequently, it is only natural that if an electric current is put through the wire from an outside source, a mechanical rotation will be induced. (Without going into details, such mechanical rotation can be brought about by both direct current and alternating current. Some motors are designed to run on either.)

You can therefore have two essentially identical devices. The first, used as a generator, will convert the heat energy of burning coal into the mechanical energy of a turning armature and convert that in turn into electrical energy. The electrical energy so produced is poured into the second device, used as a motor, and there it is converted into the mechanical energy of a turning armature. Of course, the generator can be large, supplying enough electric energy to run numerous small motors.

Once large generators made it practical to produce large quantities of electricity, and transformers made it practical to transport large quantities of electricity, it was only necessary that this electricity be conducted to the millions of motors in homes and factories.* The necessary motors were waiting to be used in this fashion; they had been waiting for some half a century, for the first practical motor had been devised by Henry in 1831.

Turning wheels had been used to supply mechanical energy as far back as earliest historic times, for rotational motion is not only useful in itself but it can easily be converted into back-and-forth motion by proper mechanical connections. Through most of man's history, wheels had been turned by the muscle of men and animals, by the action of falling water, and by blowing wind. Muscles, however, were weak and easily tired, water fell only in certain regions, and the wind was always uncertain.

After the invention of the steam engine, wheels could be turned by the application of steam power. However the bulky engines had to exist on the spot where the wheels turned and could profitably be set up only in factories or on large contrivances such as locomotives and ships. They were therefore profitably used only for large-scale work. Small-scale steam engines for home use seemed out of the question. Furthermore, it took time to start steam engines, for large quantities of water had first to be brought to a boil.

With the motor, however, an independent wheel became pos-

---

* Electricity can be put to good use without motors, of course, as in toasters and electric lights, where the heating effect alone is desired and mechanical motion is not necessary.

sible. The generator, as the source of the energy, did not have to be on the premises or anywhere near them. Moreover, electric motors could start at the flick of one switch and stop at the flick of another. Motors were versatile in the extreme, and wheels of any size and power could be turned. Huge motors were designed to move streetcars or industrial machinery, and tiny motors now power typewriters, shavers, and toothbrushes.

Thanks to Faraday and Henry (with assists from Tesla and Steinmetz), the lives of the citizens of the industrial portions of the world have thus come to be composed, in large measure, of a complex heap of electrical gadgetry.

14

Electromagnetic Radiation

Maxwell's Equations

By mid-nineteenth century, the connection between electricity and magnetism was well established and being put to good use. The generator and the motor had been invented, and both depended on the interrelationship of electricity and magnetism.

Theory lagged behind practice, however. Faraday, for instance, perhaps the greatest electrical innovator of all, was completely innocent of mathematics, and he developed his notion of lines of force in a remarkably unsophisticated way, picturing them almost like rubber bands.*

In the 1860's, Maxwell, a great admirer of Faraday, set about supplying the mathematical analysis of the interrelationship of electricity and magnetism in order to round out Faraday's non-mathematical treatment.

To describe the manner in which an electric current invariably produced a magnetic field, and in which a magnet could be made to produce an electric current, as well as how both electric charges and magnetic poles could set up fields consisting

---

* This is not meant as a sneer at Faraday, who was certainly one of the greatest scientists of all time. His intuition was that of a first-class genius. Although his views were built up without the aid of a carefully worked out mathematical analysis, they were solid. When the mathematics was finally supplied, the essence of Faraday's notions was shown to be correct.

of lines of force, in 1864 Maxwell devised a set of four comparatively simple equations,* known ever since as *Maxwell's equations*. From these, it proved possible to deduce the nature of the interrelationships of electricity and magnetism under all possible conditions.

In order for the equations to be valid, it seemed impossible to consider an electric field or a magnetic field in isolation. The two were always present together, directed at mutual right angles, so that there was a single *electromagnetic field*.

Furthermore, in considering the implications of his equations, Maxwell found that a changing electric field had to induce a changing magnetic field, which in turn had to induce a changing electric field, and so on; the two leap-frogged, so to speak, and the field progressed outward in all directions. The result was a radiation possessing the properties of a wave-form. In short, Maxwell predicted the existence of *electromagnetic waves* with frequencies equal to that in which the electromagnetic field waxed and waned.

It was even possible for Maxwell to calculate the velocity at which such an electromagnetic wave would have to move. He did this by taking into consideration the ratio of certain corresponding values in the equations describing the force between electric charges and the force between magnetic poles. This ratio turned out to have the value of 300,000 kilometers per second.

But this was equal to the velocity of light, and Maxwell could not accept that as a mere coincidence. Electromagnetic radiation was not a mere phantom of his equations but had a real existence. In fact, light must itself be an electromagnetic radiation.†

Maxwell's equations served several general functions. First, they did for the field-view of the universe what Newton's laws of motion had done for the mechanist-view of the universe.

Indeed, Maxwell's equations were more successful than Newton's laws. The latter were shown to be but approximations that held for low velocities and short distances. They required the

* Unfortunately, they are differential equations, involving the concepts of calculus, and calculus is not being used in this book. For that reason, Maxwell's equations will be spoken of, but will not be brought on stage.

† An explanation of the exact manner in which light waves, with frequencies in the hundreds of trillions, could be produced electromagnetically had to wait half a century, however—until the quantum theory (undreamed of in Maxwell's time) could be applied to the internal structure of the atom (unknown in Maxwell's time). How this was done will be described in Volume III of this book.

modification of Einstein's broader relativistic viewpoint if they were to be made to apply with complete generality. Maxwell's equations, on the other hand, survived all the changes introduced by relativity and the quantum theory; they are as valid in the light of present knowledge as they were when they were first introduced a century ago.

Secondly, Maxwell's equations, in conjunction with the later development of the quantum theory, seem at last to supply us with a satisfactory understanding of the nature of light (a question that has occupied a major portion of this volume and serves as its central question). Earlier (see page 139) I said that even granting the particle-like aspects of light there remained the wave-like aspects, and questioned what these might be. As we see now, the wave-like aspects are the oscillating values of the electro-magnetic field. The electric and magnetic components of that field are set at mutual right angles and the whole wave progresses in a direction at right angles to both.

To Maxwell, wedded to the ether hypothesis, it seemed the oscillation of the electromagnetic field consisted of wave-like distortions of the ether. However, Maxwell's equations rose superior even to Maxwell. Though the ether hypothesis passed away, the electromagnetic wave remained, for now it became possible to view the oscillating field as oscillating changes in the geometry of space. This required the presence of no matter. Nothing "had to wave" in order to form light waves.

Of the four different phenomena which, from Newton's time onward, have threatened to involve action-at-a-distance, no less than three, thanks to Maxwell's equations, were shown to be different aspects of a single phenomenon. Electricity, magnetism and light were all included in the electromagnetic field. Only the gravitational force remained outside. Maxwell, recognizing the important differences between gravity and electromagnetism, made no attempt to include the gravitational field in his equations. Since his time, some attempts have been made, notably by Einstein in the latter half of his life. Einstein's conclusions, however, have not been accepted by physicists generally, and the question of a "unified field theory" remains open.

It seemed to Maxwell that the processes that gave rise to electromagnetic radiation could well serve to produce waves of any frequency at all and not merely those of light and its near neighbors, ultraviolet and infrared radiation. He predicted, therefore, that electromagnetic radiation, in all essentials similar to light, could exist at frequencies far below and far above those of light.

Unfortunately, Maxwell did not live to see this prediction verified, for he died of cancer in 1879 at the comparatively early age of 48. Only nine years after that, in 1888, the German physicist Heinrich Rudolf Hertz (1857–1894) discovered electromagnetic radiation of very low frequency—radiation that we now call *radio waves*. This completely bore out Maxwell's prediction and was accepted as evidence for the validity of Maxwell's equations. In 1895, another German physicist, Wilhelm Konrad Röntgen (1845–1923), discovered what turned out to be electromagnetic radiation of very high frequency: radiation we now call *X rays*.

The decades of the 1880's and 1890's also saw a fundamental advance made in the study of electricity. Electric currents were driven through near-vacuums, and electrons, instead of remaining concealed in metal wires or being considered attachments to drifting atoms and groups of atoms in solution, made their appearance as particles in their own right.

The study of the new particles and radiations introduced a virtual revolution in physics and electrical technology—one so intense that it has been referred to as the Second Scientific Revolution (the First, of course, being that initiated by Galileo).

It is with the Second Scientific Revolution that the third volume of this book will deal.

Suggested Further Reading

Feather, Norman, *The Physics of Vibrations and Waves*, Edinburgh University Press, Edinburgh (1961).

Feynman, Richard P.; Leighton, Robert B.; and Sands, Matthew, *The Feynman Lectures on Physics* (Volume II), Addison-Wesley Publishing Co., Inc., Reading, Mass. (1963).

Gardner, Martin, *Relativity for the Million*, The Macmillan Co., New York (1962).

Hoffman, Banesh, *The Strange Story of the Quantum*, Dover Publications, Inc., New York (1959).

Taylor, Lloyd W., *Physics* (Volume II), Dover Publications, Inc., New York (1941).

INDEX